Peaking of World Oil Production: Impacts, Mitigation and Risk Management

Peaking of World Oil Production: Impacts, Mitigation and Risk Management

Robert L. Hirsch
Roger Bezdek
Robert Wendling

Novinka Books
New York

For permission to use material from this book please contact us:
Telephone 631-231-7269; Fax 631-231-8175
Web Site: http://www.novapublishers.com

Library of Congress Cataloging-in-Publication Data

ISBN 978-1-60021-053-2

Published by Nova Science Publishers, Inc. ✢New York

CONTENTS

PREFACE*

The peaking of world oil production presents the U.S. and the world with an unprecedented risk management problem. As peaking is approached, liquid fuel prices and price volatility will increase dramatically, and, without timely mitigation, the economic, social, and political costs will be unprecedented. Viable mitigation options exist on both the supply and demand sides, but to have substantial impact, they must be initiated more than a decade in advance of peaking.

In 2003, the world consumed just under 80 million barrels per day (MM bpd) of oil. U.S. consumption was almost 20 MM bpd, two-thirds of which was in the transportation sector. The U.S. has a fleet of about 210 million automobiles and light trucks (vans, pick-ups, and SUVs). The average age of U.S. automobiles is nine years. Under normal conditions, replacement of only half the automobile fleet will require 10-15 years. The average age of light trucks is seven years. Under normal conditions, replacement of one-half of the stock of light trucks will require 9-14 years. While significant improvements in fuel efficiency are possible in automobiles and light trucks, any affordable approach to upgrading will be inherently time-consuming, requiring more than a decade to achieve significant overall fuel efficiency improvement.

Besides further oil exploration, there are commercial options for increasing world oil supply and for the production of substitute liquid fuels: 1) Improved Oil Recovery (IOR) can marginally increase production from existing reservoirs; one of the largest of the IOR opportunities is Enhanced Oil Recovery (EOR), which can help moderate oil production declines from reservoirs that are past their peak production: 2) Heavy oil / oil sands

* Extracted from http://www.energybulletin.net/4638.html

represents a large resource of lower grade oils, now primarily produced in Canada and Venezuela; those resources are capable of significant production increases;. 3) Coal liquefaction is a well-established technique for producing clean substitute fuels from the world's abundant coal reserves; and finally, 4) Clean substitute fuels can be produced from remotely located natural gas, but exploitation must compete with the world's growing demand for liquefied natural gas. However, world-scale contributions from these options will require 10-20 years of accelerated effort.

Dealing with world oil production peaking will be extremely complex, involve literally trillions of dollars and require many years of intense effort. To explore these complexities, three alternative mitigation scenarios were analyzed:

- Scenario I assumed that action is not initiated until peaking occurs.
- Scenario II assumed that action is initiated 10 years before peaking.
- Scenario III assumed action is initiated 20 years before peaking.

For this analysis estimates of the possible contributions of each mitigation option were developed, based on an assumed crash program rate of implementation.

Our approach was simplified in order to provide transparency and promote understanding. Our estimates are approximate, but the mitigation envelope that results is believed to be directionally indicative of the realities of such an enormous undertaking. The inescapable conclusion is that more than a decade will be required for the collective contributions to produce results that significantly impact world supply and demand for liquid fuels.

Important observations and conclusions from this study are as follows:

1. When world oil peaking will occur is not known with certainty. A fundamental problem in predicting oil peaking is the poor quality of and possible political biases in world oil reserves data. Some experts believe peaking may occur soon. This study indicates that "soon" is within 20 years.
2. The problems associated with world oil production peaking will not be temporary, and past "energy crisis" experience will provide relatively little guidance. The challenge of oil peaking deserves immediate, serious attention, if risks are to be fully understood and mitigation begun on a timely basis.

3. Oil peaking will create a severe liquid fuels problem for the transportation sector, not an "energy crisis" in the usual sense that term has been used.
4. Peaking will result in dramatically higher oil prices, which will cause protracted economic hardship in the United States and the world. However, the problems are not insoluble. Timely, aggressive mitigation initiatives addressing both the supply and the demand sides of the issue will be required.
5. In the developed nations, the problems will be especially serious. In the developing nations peaking problems have the potential to be much worse.
6. Mitigation will require a minimum of a decade of intense, expensive effort, because the scale of liquid fuels mitigation is inherently extremely large.
7. While greater end-use efficiency is essential, increased efficiency alone will be neither sufficient nor timely enough to solve the problem. Production of large amounts of substitute liquid fuels will be required. A number of commercial or near-commercial substitute fuel production technologies are currently available for deployment, so the production of vast amounts of substitute liquid fuels is feasible with existing technology.
8. Intervention by governments will be required, because the economic and social implications of oil peaking would otherwise be chaotic. The experiences of the 1970s and 1980s offer important guides as to government actions that are desirable and those that are undesirable, but the process will not be easy.

Mitigating the peaking of world conventional oil production presents a classic risk management problem:

- Mitigation initiated earlier than required may turn out to be premature, if peaking is long delayed.
- If peaking is imminent, failure to initiate timely mitigation could be extremely damaging.

Prudent risk management requires the planning and implementation of mitigation well before peaking. Early mitigation will almost certainly be less expensive than delayed mitigation. A unique aspect of the world oil peaking problem is that its timing is uncertain, because of inadequate and potentially biased reserves data from elsewhere around the world. In addition, the onset of peaking may be obscured by the volatile nature of oil prices. Since the potential economic impact of peaking is immense and the uncertainties relating to all facets of the problem are large, detailed quantitative studies to address the uncertainties and to explore mitigation strategies are a critical need.

The purpose of this analysis was to identify the critical issues surrounding the occurrence and mitigation of world oil production peaking. We simplified many of the complexities in an effort to provide a transparent analysis. Nevertheless, our study is neither simple nor brief. We recognize that when oil prices escalate dramatically, there will be demand and economic impacts that will alter our simplified assumptions. Consideration of those feedbacks will be a daunting task but one that should be undertaken.

Our study required that we make a number of assumptions and estimates. We well recognize that in-depth analyses may yield different numbers. Nevertheless, this analysis clearly demonstrates that the key to mitigation of world oil production peaking will be the construction a large number of substitute fuel production facilities, coupled to significant increases in transportation fuel efficiency. The time required to mitigate world oil production peaking is measured on a decade time-scale. Related production facility size is large and capital intensive. How and when governments decide to address these challenges is yet to be determined.

Our focus on existing commercial and near-commercial mitigation technologies illustrates that a number of technologies are currently ready for immediate and extensive implementation. Our analysis was not meant to be limiting. We believe that future research will provide additional mitigation options, some possibly superior to those we considered. Indeed, it would be appropriate to greatly accelerate public and private oil peaking mitigation research. However, the reader must recognize that doing the research required to bring new technologies to commercial readiness takes time under the best of circumstances. Thereafter, more than a decade of intense implementation will be required for world scale impact, because of the inherently large scale of world oil consumption.

In summary, the problem of the peaking of world conventional oil production is unlike any yet faced by modern industrial society. The challenges and uncertainties need to be much better understood. Technologies exist to mitigate the problem. Timely, aggressive risk management will be essential.

ACKNOWLEDGEMENTS

This work was sponsored by the National Energy Technology Laboratory of the Department of Energy, under Contracts No. DE-AM26-99FT40575, Task 21006W and Subcontract Agreement number 7010001197 with Energy and Environmental Solutions, LLC. The authors are indebted to NETL management for their encouragement and support.

INTRODUCTION

Oil is the lifeblood of modern civilization. It fuels the vast majority of the world's mechanized transportation equipment – Automobiles, trucks, airplanes, trains, ships, farm equipment, the military, etc. Oil is also the primary feedstock for many of the chemicals that are essential to modern life. This study deals with the upcoming physical shortage of world conventional oil --an event that has the potential to inflict disruptions and hardships on the economies of every country.

The earth's endowment of oil is finite and demand for oil continues to increase with time. Accordingly, geologists know that at some future date, conventional oil supply will no longer be capable of satisfying world demand. At that point world conventional oil production will have peaked and begin to decline.

Table 1. Predictions of World Oil Production Peaking

Projected Date	Source of Projection
2006-2007	Bakhitari
2007-2009	Simmons
After 2007	Skrebowski
Before 2009	Deffeyes
Before 2010	Goodstein
Around 2010	Campbell
After 2010	World Energy Council
2010-2020	Laherrere
2016	EIA (Nominal)
After 2020	CERA
2025 or later	Shell
No visible Peak	Lynch

A number of experts project that world production of conventional oil could occur in the relatively near future, as summarized in Table I-1.[1] Such projections are fraught with uncertainties because of poor data, political and institutional self-interest, and other complicating factors. The bottom line is that no one knows with certainty when world oil production will reach a peak,[2] but geologists have no doubt that it will happen.

Our aim in this study is to:

- Summarize the difficulties of oil production forecasting;
- Identify the fundamentals that show why world oil production peaking is such a unique challenge;
- Show why mitigation will take a decade or more of intense effort;
- Examine the potential economic effects of oil peaking;
- Describe what might be accomplished under three example mitigation scenarios.
- Stimulate serious discussion of the problem, suggest more definitive studies, and engender interest in timely action to mitigate its impacts.

In Chapter II we describe the basics of oil production, the meaning of world conventional oil production peaking, the challenge of making accurate forecasts, and the effects that higher prices and advanced technology might have on oil production.

Because of the massive scale of oil use around the world, mitigation of oil shortages will be difficult, time consuming, and expensive. In Chapter III we describe the extensive and critical uses of U.S. oil and the long economic and mechanical lifetimes of existing liquid fuel consuming vehicles and equipment.

While it is impossible to predict the impact of world oil production peaking with any certainty, much can be learned from past oil disruptions, particularly the 1973 oil embargo and the 1979 Iranian oil shortage, as discussed in Chapter IV. In Chapter V we describe the developing shortages of U.S. natural gas, shortages that are occurring in spite of assurances of abundant supply provided just a few years ago. The parallels to world oil supply are disconcerting.

In Chapter VI we describe available mitigation options and related implementation issues. We limit our considerations to technologies that are near ready or currently commercially available for immediate deployment. Clearly, accelerated research and development holds promise for other options. However, the challenge related to extensive near-term oil shortages will require deployment of currently viable technologies, which is our focus.

Oil is a commodity found in over 90 countries, consumed in all countries, and traded on world markets. To illustrate and bracket the range of mitigation **options**, we developed three illustrative scenarios. Two assume action well in advance of the onset of world oil peaking – in one case, 20 years before peaking and in another case, 10 years in advance. Our third scenario assumes that no action is taken prior to the onset of peaking. Our findings illustrate the magnitude of the problem and the importance of prudent risk management.

Finally, we touch on possible market signals that might foretell the onset of peaking and possible wildcards that might change the timing of world conventional oil production peaking. In conclusion, we frame the challenge of an unknown date for peaking, its potentially extensive economic impacts, and available mitigation options as a matter of risk management and prudent response. The reader is asked to contemplate three major questions:

- What are the risks of heavy reliance on optimistic world oil production peaking projections?
- Must we wait for the onset of oil shortages before actions are taken?
- What can be done to ensure that prudent mitigation is initiated on a timely basis?

Chapter I

PEAKING OF WORLD OIL PRODUCTION [3]

A. BACKGROUND

Oil was formed by geological processes millions of years ago and is typically found in underground reservoirs of dramatically different sizes, at varying depths, and with widely varying characteristics. The largest oil reservoirs are called "Super Giants," many of which were discovered in the Middle East. Because of their size and other characteristics, Super Giant reservoirs are generally the easiest to find, the most economic to develop, and the longest lived. The last Super Giant oil reservoirs discovered worldwide were found in 1967 and 1968. Since then, smaller reservoirs of varying sizes have been discovered in what are called "oil prone" locations worldwide --oil is not found everywhere.

Geologists understand that oil is a finite resource in the earth's crust, and at some future date, world oil production will reach a maximum --a peak --after which production will decline. This logic follows from the well-established fact that the output of individual oil reservoirs rises after discovery, reaches a peak and declines thereafter. Oil reservoirs have lifetimes typically measured in decades, and peak production often occurs roughly a decade or so after discovery. It is important to recognize that *oil production peaking is not "running out."* Peaking is a reservoir's maximum oil production rate, which typically occurs after roughly half of the recoverable oil in a reservoir has been produced. In many ways, what is likely to happen on a world scale is similar to what happens to individual reservoirs, because world production is the sum total of production from many different reservoirs.

Because oil is usually found thousands of feet below the surface and because oil reservoirs normally do not have an obvious surface signature, oil is very difficult to find. Advancing technology has greatly improved the discovery process and reduced exploration failures. Nevertheless, oil exploration is still inexact and expensive.

Once oil has been discovered via an exploratory well, full-scale production requires many more wells across the reservoir to provide multiple paths that facilitate the flow of oil to the surface. This multitude of wells also helps to define the total recoverable oil in a reservoir – its so-called "reserves."

B. Oil Reserves

The concept of reserves is generally not well understood. "Reserves" is an estimate of the amount of oil in a reservoir that can be extracted at an assumed cost. Thus, a higher oil price outlook often means that more oil can be produced, but geology places an upper limit on price-dependent reserves growth; in well managed oil fields, it is often 10-20 percent more than what is available at lower prices.

Reserves estimates are revised periodically as a reservoir is developed and new information provides a basis for refinement. Reserves estimation is a matter of gauging how much extractable oil resides in complex rock formations that exist typically one to three miles below the surface of the ground, using inherently limited information. Reserves estimation is a bit like a blindfolded person trying to judge what the whole elephant looks like from touching it in just a few places. It is not like counting cars in a parking lot, where all the cars are in full view.

Specialists who estimate reserves use an array of methodologies and a great deal of judgment. Thus, different estimators might calculate different reserves from the same data. Sometimes politics or self-interest influences reserves estimates, e.g., an oil reservoir owner may want a higher estimate in order to attract outside investment or to influence other producers.

Reserves and production should not be confused. Reserves estimates are but one factor in estimating future oil production from a given reservoir. Other factors include production history, understanding of local geology, available technology, oil prices, etc. An oil field can have large estimated reserves, but if the field is past its maximum production, the remaining reserves will be produced at a declining rate. This concept is important because *satisfying increasing oil demand not only requires continuing to produce older oil reservoirs with their declining production, it also requires finding new ones, capable of producing sufficient quantities of oil to both compensate for shrinking production from older fields and to provide the increases demanded by the market.*

C. Production Peaking

World oil demand is expected to grow 50 percent by 2025.[4] To meet that demand, ever-larger volumes of oil will have to be produced. Since oil production from individual reservoirs grows to a peak and then declines, new reservoirs must be continually discovered and brought into production to compensate for the depletion of older reservoirs. If large quantities of new oil are not discovered and brought into production somewhere in the world, then world oil production will no longer satisfy demand. That point is called the peaking of world conventional oil production.

When world oil production peaks, there will still be large reserves remaining. Peaking means that the rate of world oil production cannot increase; it also means that production will thereafter decrease with time.

The peaking of world oil production has been a matter of speculation from the beginning of the modern oil era in the mid 1800s. In the early days, little was known about petroleum geology, so predictions of peaking were no more than guesses without basis. Over time, geological understanding improved dramatically and guessing gave way to more informed projections, although the knowledge base involves numerous uncertainties even today.

Past predictions typically fixed peaking in the succeeding 10-20 year period. Most such predictions were wrong, which does not negate that peaking will someday occur. Obviously, we cannot know if recent forecasts are wrong until predicted dates of peaking pass without incident.

With a history of failed forecasts, why revisit the issue now? The reasons are as follows:

1. Extensive drilling for oil and gas has provided a massive worldwide database; current geological knowledge is much more extensive than in years past, i.e., we have the knowledge to make much better estimates than previously.
2. Seismic and other exploration technologies have advanced dramatically in recent decades, greatly improving our ability to

discover new oil reservoirs. Nevertheless, the oil reserves discovered per exploratory well began dropping worldwide over a decade ago. We are finding less and less oil in spite of vigorous efforts, suggesting that nature may not have much more to provide.

3. Many credible analysts have recently become much more pessimistic about the possibility of finding the huge new reserves needed to meet growing world demand.
4. Even the most optimistic forecasts suggest that world oil peaking will occur in less than 25 years.
5. The peaking of world oil production could create enormous economic disruption, as only glimpsed during the 1973 oil embargo and the 1979 Iranian oil cut-off.

Accordingly, there are compelling reasons for in-depth, unbiased reconsideration.

D. Types of Oil

Oil is classified as "Conventional" and "Unconventional." Conventional oil is typically the highest quality, lightest oil, which flows from underground reservoirs with comparative ease. Unconventional oils are heavy, often tar-like. They are not readily recovered since production typically requires a great deal of capital investment and supplemental energy in various forms. For that reason, most current world oil production is conventional oil.[5] (Unconventional oil production will be discussed in Chapter VI).

E. Oil Resources[6]

Consider the world resource of conventional oil. In the past, higher prices led to increased estimates of conventional oil reserves worldwide. However, this price-reserves relationship has its limits, because *oil is found in discrete packages (reservoirs) as opposed to the varying concentrations characteristic of many minerals.* Thus, at some price, world reserves of recoverable conventional oil will reach a maximum because of geological fundamentals. Beyond that point, insufficient additional conventional oil will be recoverable at any realistic price. This is a geological fact that is often misunderstood by people accustomed to dealing with hard minerals, whose geology is fundamentally different. This misunderstanding often clouds rational discussion of oil peaking.

Future world recoverable reserves are the sum of the oil remaining in existing reservoirs plus the reserves to be added by future oil discoveries. *Future oil production* will be the sum of production from older reservoirs in decline, newer reservoirs from which production is increasing, and yet-to-be discovered reservoirs.

Because oil prices have been relatively high for the past decade, oil companies have conducted extensive exploration over that period, but their results have been disappointing. If recent trends hold, there is little reason to expect that exploration success will dramatically improve in the future. This situation is evident in Figure I.1, which shows the difference between annual world oil reserves additions minus annual consumption.[7] The image is one of a world moving from a long period in which reserves additions were much greater than consumption, to an era in which annual additions are falling increasingly short of annual consumption. This is but one of a number of trends that suggest the world is fast approaching the inevitable peaking of conventional world oil production.

F. Impact of Higher Prices and New Technology

Conventional oil has been the mainstay of modern civilization for more than a century, because it is most easily brought to the surface from deep underground reservoirs, and it is the most easily refined into finished fuels. The U.S. was endowed with huge reserves of petroleum, which underpinned U.S. economic growth in the early and mid twentieth century. However, U.S. oil resources, like those in the world, are finite, and growing U.S. demand resulted in the peaking of U.S. oil production in the Lower 48 states in the early 1970s. With relatively minor exceptions, U.S. Lower 48 oil production has been in continuing decline ever since. Because U.S. demand for petroleum products continued to increase, the U.S. became an oil importer. Today, the U.S. depends on foreign sources for almost 60 percent of its needs, and future U.S. imports are projected to rise to 70 percent of demand by 2025.[8]

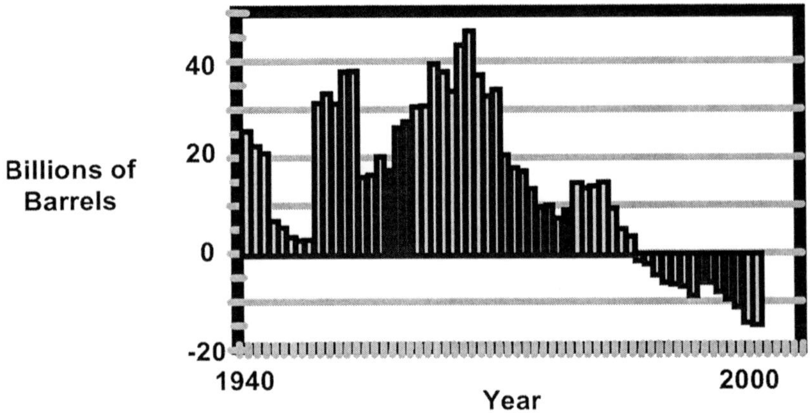

Figure I.1. Net Difference Between Annual World Oil Reserves Additions and Annual Consumption

Over the past 50 years, exploration for and production of petroleum has been an increasingly more technological enterprise, benefiting from more sophisticated engineering capabilities, advanced geological understanding, improved instrumentation, greatly expanded computing power, more durable materials, etc. Today's technology allows oil reservoirs to be more readily discovered and better understood sooner than heretofore. Accordingly, reservoirs can be produced more rapidly, which provides significant economic advantages to the operators but also hastens peaking and depletion.

Some economists expect higher oil prices and improved technologies to continue to provide ever-increasing oil production for the foreseeable future. Most geologists disagree because they do not believe that there are many huge new oil reservoirs left to be found. Accordingly, geologists and other observers believe that supply will eventually fall short of growing world demand – and result in the peaking of world conventional oil production.

To gain some insight into the effects of higher oil prices and improved technology on oil production, let us briefly examine related impacts in the U.S. Lower 48 states. This region is a useful surrogate for the world, because it was one of the world's richest, most geologically varied, and most productive up until 1970, when production peaked and started into decline. While the U.S. is the best available surrogate, it should be remembered that the decline rate in US production was in part impacted by the availability of large volumes of relatively low cost oil from the Middle East.

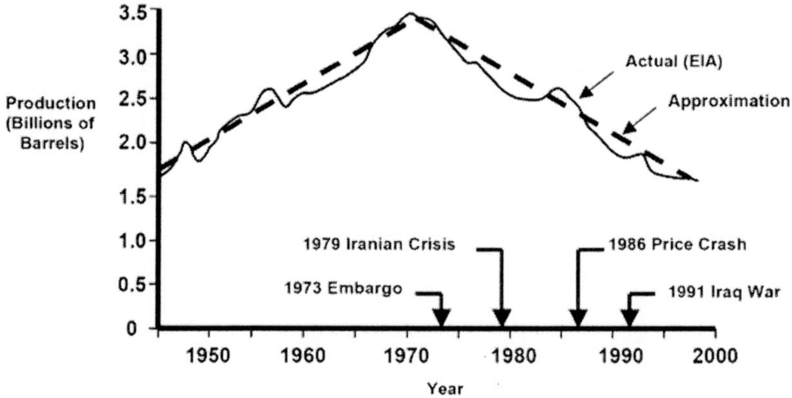

Figure I.2. U.S. Lower 48 Oil Production, 1945-2000

Figure I.2 shows EIA data for Lower 48 oil production,[9] to which trend lines have been added that will aid our scenarios analysis later in the report. The trend lines show a relatively symmetric, triangular pattern. For reference, four notable petroleum market events are noted in the figure: the 1973 OPEC oil embargo, the 1979 Iranian oil crisis, the 1986 oil price collapse, and the 1991 Iraq war.

Figure I.3 shows Lower 48 historical oil production with oil prices and technology trends added. In constant dollars, oil prices increased by roughly a factor of three in 1973-74 and another factor of two in 1979-80. The modest production up-ticks in the mid 1980s and early 1990s are likely responses to the 1973 and 1979 oil price spikes, both of which spurred a major increase in U.S exploration and production investments. The delays in production response are inherent to the implementation of large-scale oil field investments. The fact that the production up-ticks were moderate was due to the absence of attractive exploration and production opportunities, because of geological realities. Beyond oil price increases, the 1980s and 1990s were a golden age of oil field technology development, including practical 3-D seismic, economic horizontal drilling, and dramatically improved geological understanding. Nevertheless, as Figure I.3 shows, Lower 48 production still trended downward, showing no pronounced response to either price or technology. In light of this experience, there is good reason to expect that an analogous situation will exist worldwide after world oil production peaks: *Higher prices and improved technology are unlikely to yield dramatically higher conventional oil production.[*10]

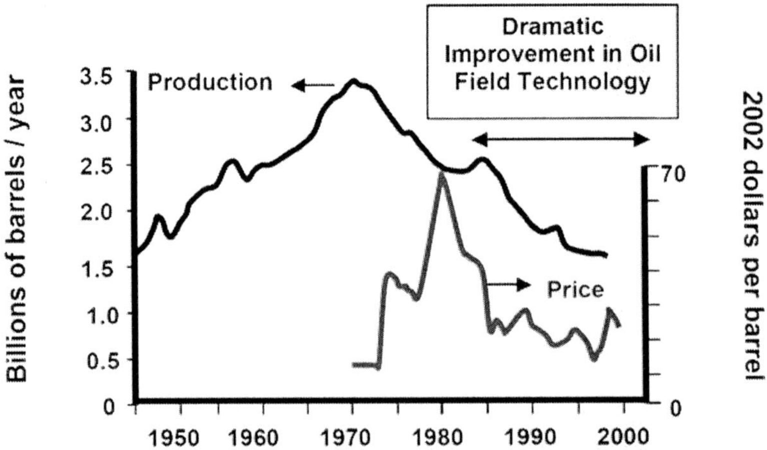

Figure I.3. Lower 48 Oil Production and Oil Prices

G. PROJECTIONS OF THE PEAKING OF WORLD OIL PRODUCTION

Table I.1. Projections of the Peaking of World Oil Production

Projected Date	*Source of Projection*	*Background & Reference*
2006-2007	Bakhitari, A.M.S.	Iranian Oil Executive[11]
2007-2009	Simmons, M.R.	Investment banker [12]
After 2007	Skrebowski, C.	Petroleum journal Editor [13]
Before 2009	Deffeyes, K.S.	Oil company geologist (ret.) [14]
Before 2010	Goodstein, D.	Vice Provost, Cal Tech [15]
Around 2010	Campbell, C.J.	Oil company geologist (ret.) [16]
After 2010	World Energy Council	World Non-Government Org. [17]
2010-2020	Laherrere, J.	Oil company geologist (ret.) [18]
2016	EIA nominal case	DOE analysis/ information [19]
After 2020	CERA	Energy consultants [20]
2025 or later	Shell	Major oil company [21]
No visible peak	Lynch, M.C.	Energy economist [22]

Projections of future world oil production will be the sum total of 1) output from all of the world's then existing producing oil reservoirs, which will be in various stages of development, and 2) all the yet-to-be discovered reservoirs in their various states of development. This is an extremely complex summation problem, because of the variability and possible biases in publicly available data. In practice, estimators use various approximations to predict future world oil production. The remarkable complexity of the problem can easily lead to incorrect conclusions, either positive or negative.

Various individuals and groups have used available information and geological estimates to develop projections for when world oil production might peak. A sampling of recent projections is shown in Table I.1.

WHY THE TRANSITION WILL BE SO TIME CONSUMING

A. Introduction

Use of petroleum is pervasive throughout the U.S. economy. It is directly linked to all market sectors because all depend on oil-consuming capital stock. Oil price shocks and supply constraints can often be mitigated by temporary decreases in consumption; however, long term price increases resulting from oil peaking will cause more serious impacts. Here we examine historical oil usage patterns by market sector, provide a summary of current consumption patterns, identify the most important markets, examine the relationship between oil and capital stock, and provide estimates of the time and costs required to transition to more energy efficient technologies that can play a role in mitigating the adverse effects of world oil peaking.

B. Historical U.S. Oil Consumption Patterns

After the two oil price shocks and supply disruptions in 1973-74 and 1979, oil consumption in the U.S. decreased 13 percent, declining from nearly 35 quads in 1973 to 30 quads in 1983. However, overall consumption continued to grow after the 1983 low and has continuously increased over the last 20 years, reaching over 39 quads in 2003, as shown in Figure I.1. Of particular note are changes in three U.S. market sectors: 1) Oil consumption in the residential sector declined from eight percent of total oil consumption in 1973 to four percent in 2003, a decrease of 50 percent; 2) Oil consumption in the commercial sector declined from five percent to two percent, decreasing 58 percent; and 3) Consumption in the electric power sector fell from 10 percent in 1973 to three percent in 2003, decreasing 70 percent. These three market sectors currently account for 1.3 quads of oil consumption annually, representing nine percent of U.S. oil demand in 2003.

Oil consumption in other market sectors did not decrease. A 140 percent growth in GDP over the 1973-2003 period made it difficult to decrease oil consumption in the industrial and transportation sectors.[23] In particular, personal transportation grew significantly over the past three decades, and total vehicle miles traveled for cars and light trucks more than doubled over the period.[24] From 1973 to 2003, consumption of oil in the industrial sector stayed relatively flat at just over nine quads, and the industrial sector's share of total U.S. consumption remained between 24 and 26 percent. In sharp contrast to all other sectors, U.S. oil consumption for transportation purposes has increased steadily every year, rising from just over 17 quads in 1973 to 26 quads in 2003. By 2003, the transportation sector accounted for two-thirds of the oil consumed in the U.S.

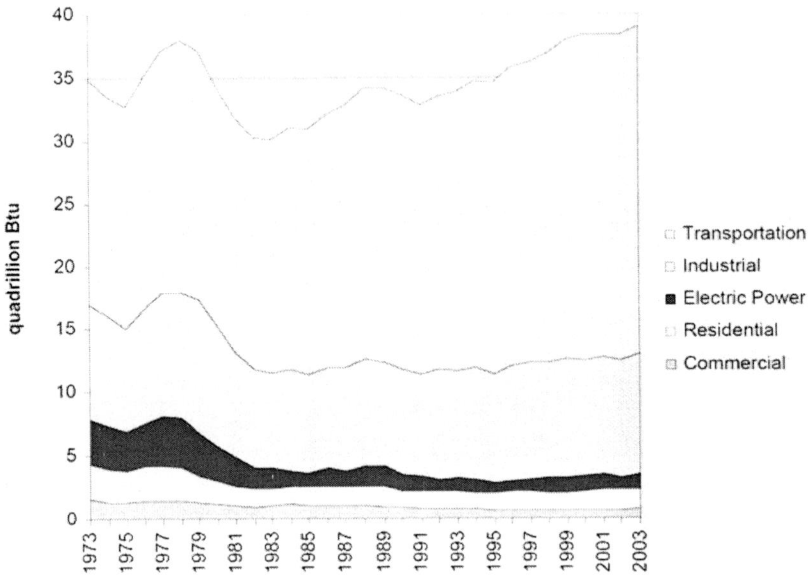

Figure I.4. U.S. Petroleum Consumption by Sector, 1973-2003 [25]

C. Petroleum in the Current U.S. Economy

The 39 quad consumption of oil in the U.S. in 2003 is equivalent to 19.7 million barrels of oil per day (MM bpd), including almost 13.1 MM bpd consumed by the transportation sector and 4.9 MM bpd by the industrial sector, as shown in Table I.2. This table also shows the petroleum fuel types consumed by each sector. Motor gasoline consumption accounted for 45 percent of U.S. daily petroleum consumption, nearly 9 MM bpd, almost all of which was used in autos and light trucks. Distillate fuel oil was the second-most consumed oil product at almost 3.8 MM bpd (19 percent of consumption), and most was used as diesel fuel for medium and heavy trucks. Finally, the third most consumed oil product was liquefied petroleum gases, at 2.2 MM bpd equivalent (11 percent of total consumption), most of which was used in the industrial sector as feedstock by the chemicals industry. Only two other consuming areas exceeded the 1 MM bpd level: kerosene and jet fuel in the transportation sector, primarily for airplanes, and "other petroleum" by the industrial sector, primarily petroleum feedstocks used to produce non-fuel products in the petroleum and chemical industries.

Table I.2.Detailed Consumption of Petroleum in the U.S.by Fuel Type and Sector –2003[26] (Thousand of barrels per day)

	Residential	Commercial	Industrial	Transportation	Electric Power	Total
Motor Gasoline	-	20	159	8,665	-	8,844
Distillate Fuel Oil	421	236	603	2,455	51	3,766
LPG	429	76	1,648	10	-	2,163
Kerosene/Jet Fuel	27	9	7	1,608	-	1,651
Residual	-	30	87	250	291	658
Asphalt & Road Oil	-	-	513	-	-	513
Petroleum Coke	-	-	398	-	61	459
Lubricants	-	-	78	73	-	151
Aviation Gas	-	-	-	18	-	18
Other Petroleum	-	-	1,435	-	-	1,435
Total	877	371	4,928	13,079	403	19,658

D. Capital Stock Characteristics in the Largest Consuming Sectors

Energy efficiency improvements and technological changes are typically incorporated into products and services slowly, and their rate of market penetration is based on customer preferences and costs. In the 1974-1983 period, oil prices ratcheted up to newer, higher levels, which lead to significant energy efficiency improvements, energy fuel switching, and other more general technological changes. Some changes came about due to legislative mandates (corporate average fuel economy standards, CAFE) or subsidies (solar energy and energy efficiency tax credits), but many were the result of economic decisions to reduce long-term costs. Under a normal course of replacement based on historical trends, oil-consuming capital stock has been replaced in the

U.S. over a period of 15 to 50 years and has cost consumers and businesses trillions of dollars, as discussed below.

Automobiles represent the largest single oil-consuming capital stock in the U.S. 130 million autos consume 4.9 MM bpd, or 25 percent of total consumption, as shown in Table I.3. Autos remain in the U.S. transportation fleet, or rolling stock, for a long time. While the financial-based current-cost, average age of autos is only 3.4 years, the average age of the stock is currently nine years.

Recent studies show that *one half of the1990-model year cars will remain on the road 17 years later in 2007.* At normal replacement rates, *consumers will spend* an estimated $1.3 trillion (constant 2003 dollars) over the next 10-15 years just to *replace one-half the stock of automobiles.*[27]

Table I.3.
U.S. Capital Stock Profiles

	Autos	**Light Trucks**	**Heavy Trucks**	**Air Carrier s**
Oil consumption (MM bpd)[28]	4.9	3.6	3.0	1.1
Share of the U.S. total	*25%*	*18%*	*16%*	*6%*
Current cost of net capital stock (billion $)[29]	$571 B	$435 B	$686 B	$110 B
Fleet size [30]	130 MM	80 MM	7 MM	8,500
Number of annual purchases	8.5 MM	8.5 MM	500,000	400
Average age of stock (years)	9	7	9	13
Median lifetime (years)	17	16	28	22

A similar situation exists with light trucks (vans, pick-ups, and SUVs), which consume 3.6 MM bpd of oil, accounting for 18 percent of total oil consumption. Light trucks are depreciated on a faster schedule, and their financial-based current-cost average age is 2.9 years. However, the average physical age of the rolling stock is seven years, and the median lifetime of light trucks is 16 years. At current replacement rates, *one-half of the 80-million light trucks will be replaced in the next 9-14 years at a cost of $1 trillion.*

Seven million heavy trucks (including buses, highway trucks, and off-highway trucks) represent the third largest consumer of oil at 3.0 MM bpd, 16 percent of total consumption. The current-cost average age of heavy trucks is 5.0 years, but the median lifetime of this equipment is 28 years. The disparity in the average age and the median lifetime estimates indicate that a significant number of vehicles are 40-60 years old. At normal replacement levels, *one-half of the heavy truck stock will be replaced by businesses in the next 15-20 years at a cost of $1.5 trillion.*

The fourth-largest consumer of oil is the airlines, which consume the equivalent of 1.1 MM bpd, representing six percent of U.S. consumption. The 8,500 aircraft have a current-cost average age of 9.1 years, and a median lifetime of 22 years. Airline deregulation and the events of September 11, 2001, have had significant effects on the industry, its ownership, and recent business decisions. At recent rates, *airlines will replace one-half of their stock over the next 15-20 years at a cost of $250 billion.*

These four capital stock categories cover most transportation modes and represent 65 percent of the consumption of oil in the U.S.[31] The three largest categories of autos, light trucks, and heavy trucks all utilize the internal combustion engine, whether gasoline-or diesel-burning. Clearly, advancements in energy efficiency and replacement in this capital stock (for instance, electric-hybrid engines) would help mitigate the economic impacts of rising oil prices caused by world oil peaking. However, as described, the normal replacement rates of this equipment will require 10-20 years and cost trillions of dollars. *We cannot conceive of any affordable government-sponsored "crash program" to accelerate normal replacement schedules so as to incorporate higher energy efficiency technologies into the privately-owned transportation sector; significant improvements in energy efficiency will thus be inherently time-consuming (of the order of a decade or more).*

When oil prices increase associated with oil peaking, consumers and businesses will attempt to reduce their exposure by substitution or by decreases in consumption. In the short run, there may be interest in the substitution of natural gas for oil in some applications, but the current outlook for natural gas availability and price is cloudy for a decade or more. An increase in demand for electricity in rail transportation would increase the need for more electric power plants. In the short run, much of the burden of adjustment will likely be borne by decreases in consumption from discretionary decisions, since 67 percent of personal automobile travel and nearly 50 percent of airplane travel are discretionary.[32]

E. Consumption Outside the U.S.

Oil consumption patterns differ in other countries. While two-thirds of U.S. oil use is in the transportation sector, worldwide that share is estimated about 55 percent. However, that difference is narrowing as world economic development is expanding transportation demands at an even faster pace. A portion of non-transportation oil consumption is switchable. As stated by EIA, "Oil's importance in other end-use sectors is likely to decline where other fuels are competitive, such as natural gas, coal, and nuclear, in the electric sector, but currently there is no alternative energy sources that compete economically with oil in the transportation sector."[33] Because sector-by-sector oil consumption data for many counties is unavailable, a detailed analysis of world consumption was beyond the scope of this report. Nevertheless, it is clear that transportation is the primary market for oil worldwide.

F. Transition Conclusions

Any transition of liquid fueled, end-use equipment following oil peaking will be time consuming. The depreciated value of existing U.S. transportation capital stock is nearly $2 trillion and would normally require 25 – 30 years to replace. At that rate, significantly more energy efficient equipment will only be slowly phased into the marketplace as new capital stock gradually replaces existing stock. Oil peaking will likely accelerate replacement rates, but the transition will still require decades and cost trillions of dollars.

Chapter II

LESSONS FROM PAST EXPERIENCE

A. PREVIOUS OIL SUPPLY SHORTFALL AND DISRUPTIONS

There have been over a dozen global oil supply disruptions [34] over the past half-century, as summarized in Figure II.1.

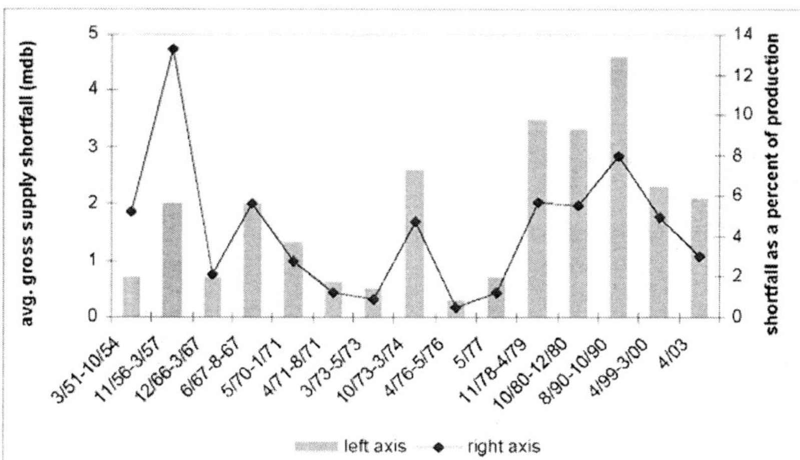

Figure II.1. Global Oil Supply Disruptions: 1954-2003

Briefly,

- Disruptions ranged in duration from one to 44 months. Supply shortfalls were 0.3 -4.6 MM bpd, and eight resulted in average gross supply shortfalls of at least 2 MM bpd.
- Percentage supply shortfalls varied from roughly one percent to nearly 14 percent of world production.
- The most traumatic disruption, 1973-74, was not the most severe, but it nevertheless lead to greatly increased oil prices and significant worldwide economic damage.
- The second most traumatic disruption, 1979, was also neither the longest nor the most severe.

For purposes of this study, the 1973-74 and 1979 disruptions are taken as the most relevant, because they are believed to offer the best insights into what might occur when world oil production peaks.

B. Difficulties in Deriving Implications from Past Experience

Over the past 30 years, most economic studies of the impact of oil supply disruptions assumed that the interruptions were temporary and that each situation would shortly return to "normal." Thus, the major focus of most studies was determination of the appropriate fiscal and monetary policies required to minimize negative economic impacts and the development of policies to help the economy and labor market adjust until the disruption ended.[35] Few economists considered a situation where the oil supply shortfall may be long-lived (a decade or more).

Since 1970, most large oil price increases were eventually followed by oil price declines, and, since these cycles were expected to be repeated, it was generally felt that "the problem will take care of itself as long at the government does nothing and does not interfere."[36] The frequent and incorrect predictions of oil shortfalls have been often used to discredit future predictions of a longer-term problem and to discredit the need for appropriate long-term U.S. energy policies.

C. How Oil Supply Shortfalls Affect the Global Economy

Oil prices play a key role in the global economy, since the major impact of an oil supply disruption is higher oil prices.[37] Oil price increases transfer income from oil importing to oil exporting countries, and the net impact on world economic growth is negative. For oil importing countries, increased oil prices reduce national income because spending on oil rises, and there is less available to spend on other goods and services.[38] Not surprisingly, the larger the oil price increase and the longer higher prices are sustained, the more severe is the macroeconomic impact.

Higher oil prices result in increased costs for the production of goods and services, as well as inflation, unemployment, reduced demand for products other than oil, and lower capital investment. Tax revenues decline and budget deficits increase, driving up interest rates. These effects will be greater the more abrupt and severe the oil price increase and will be exacerbated by the impact on consumer and business confidence.

Government policies cannot eliminate the adverse impacts of sudden, severe oil disruptions, but they can minimize them. On the other hand, contradictory monetary and fiscal policies to control inflation can exacerbate recessionary income and unemployment effects. (See Appendix II for further discussion of past government actions).

D. The U.S. Experience

As illustrated in Figure II.2, oil price increases have preceded most U.S. recessions since 1969, and virtually every serious oil price shock was followed by a recession. Thus, *while oil price spikes may not be necessary to trigger a recession in the U.S., they have proven to be sufficient over the past 30 years.*

E. The Experience of Other Countries

1. The Developed (OECD) Economies

Estimates of the damage caused by past oil price disruptions vary substantially, but without a doubt, the effects were significant. Economic growth decreased in most oil importing countries following the disruptions of 1973-74 and 1979-80, and the impact of the first oil shock was accentuated by inappropriate policy responses.[39] Despite a decline in the ratio of oil consumption to GDP over the past three decades, oil remains vital, and there is considerable empirical evidence regarding the effects of oil price shocks:

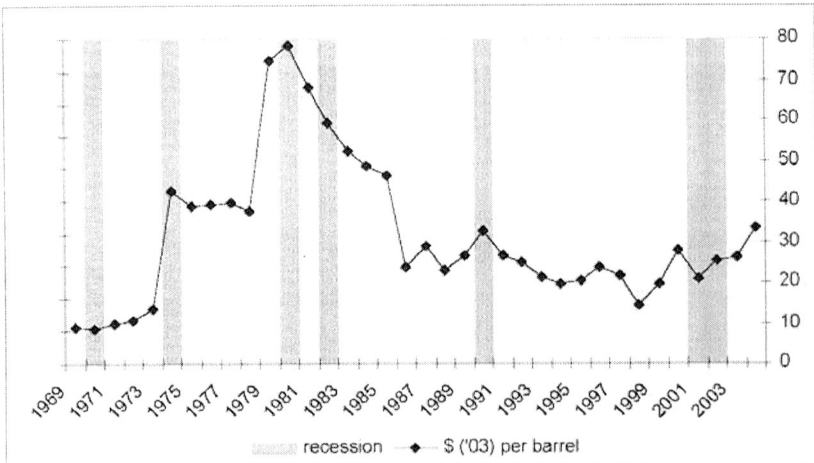

Figure II.2. Oil Prices and U.S. Recessions: 1969-2003[40]

- The loss suffered by the OECD countries in the 1974/-75 recession amounted to $350 billion (current dollars) / $1.1 trillion 2003 dollars, although part of this loss was related to factors other than oil price.[41]
- The loss resulting from the 1979 oil disruption was about three percent of GDP ($350 billion in current dollars) in 1980 rising to 4.25 percent ($570 billion) in 1981, and accounted for much of the decline in economic growth and the increase in inflation and unemployment in the OECD in 1981-82.[42]
- The effect of the 1990-91 oil price upsurge was more modest, because price increases were smaller; they did not persist; and oil intensity in OECD countries had declined.
- Although oil intensity and the share of oil in total imports have declined in recent years, OECD economies remain vulnerable to

higher oil prices, because of the "life blood" nature of liquid fuel use.

2. Developing Countries

Developing countries suffer more than the developed countries from oil price increases because they generally use energy less efficiently and because energy-intensive manufacturing accounts for a larger share of their GDP. On average, developing countries use more than twice as much oil to produce a unit of output as developed countries, and oil intensity is increasing in developing countries as commercial fuels replace traditional fuels and industrialization/urbanization continues.[43]

The vulnerability of developing countries is exacerbated by their limited ability to switch to alternative fuels. In addition, an increase in oil import costs also can destabilize trade balances and increase inflation more in developing countries, where financial institutions and monetary authorities are often relatively unsophisticated. This problem is most pronounced for the poorest developing countries.

F. IMPLICATIONS

1. The World Economy

A shortfall of oil supplies caused by world conventional oil production peaking will sharply increase oil prices and oil price volatility. As oil peaking is approached, relatively minor events will likely have more pronounced impacts on oil prices and futures markets.

Oil prices remain a key determinant of global economic performance, and world economic growth over the past 50 years has been negatively impacted in the wake of increased oil prices. The greater the supply shortfall, the higher the price increases; the longer the shortfall, the greater will be the adverse economic affects.

The long-run impact of sustained, significantly increased oil prices associated with oil peaking will be severe. Virtually certain are increases in inflation and unemployment, declines in the output of goods and services, and a degradation of living standards. Without timely mitigation, the long-run impact on the developed economies will almost certainly be extremely damaging, while many developing nations will likely be even worse off.[44]

The impact of oil price changes will likely be asymmetric. The negative economic effects of oil price increases are usually not offset by the economic stimulus resulting from a fall in oil prices. The increase in economic growth in oil exporting countries provided by higher oil prices has been less than the loss of economic growth in importing countries, and these effects will likely continue in the future.[45]

2. The United States

For the U.S., each 50 percent sustained increase in the price of oil will lower real U.S. GDP by about 0.5 percent, and a doubling of oil prices would reduce GDP by a full percentage point. Depending on the U.S. economic growth rate at the time, this could be a sufficient negative impact to drive the country into recession. Thus, assuming an oil price in the $25 per barrel range --the 2002-2003 average, an increase of the price of oil to $50 per barrel would cost the economy a reduction in GDP of around $125 billion.

If the shortfall persisted or worsened (as is likely in the case of peaking), the economic impacts would be much greater. Oil supply disruptions over the past three decades have cost the U.S. economy about $4 trillion, so supply shortfalls associated with the approach of peaking could cost the U.S. as much as all of the oil supply disruptions since the early 1970s combined.

The effects of oil shortages on the U.S. are also likely to be asymmetric. Oil supply disruptions and oil price increases reduce economic activity, but oil price declines have a less beneficial impact.[46] Oil shortfalls and price increases will cause larger responses in job destruction than job creation, and many more jobs may be lost in response to oil price increases than will be regained if oil prices were to decrease. These effects will be more pronounced when oil price volatility increases as peaking is approached. The repeated economic and job losses experienced during price spikes will not be replaced as prices decrease. As these cycles continue, the net economic and job losses will increase.

Sectoral shifts will likely be pronounced. Even moderate oil disruptions could cause shifts among sectors and industries of ten percent or more of the labor force.[47] Continuing oil shortages will likely have disruptive inter-sectoral, inter-industry, and inter-regional effects, and the sectors that are (both directly and indirectly) oil-dependant could be severely impacted.[48]

Monetary policy is more effective in controlling the inflationary effects of a supply disruption than in averting related recessionary effects.[49] Thus, while appropriate monetary policy may be successful in lessening the inflationary impacts of oil price increases, it may do so at the cost of recession and increased unemployment. Monetary policies tend to be used to increase interest rates to control inflation, and it is the high interest rates that cause most of the economic damage. As peaking is approached, devising appropriate offsetting fiscal, monetary, and energy policies will become more difficult. Economically, the decade following peaking may resemble the 1970s, only worse, with dramatic increases in inflation, long-term recession, high unemployment, and declining living standards.[50]

Chapter III

Learning From Natural Gas

A. Introduction

A dramatic example of the risks of over-reliance on geological resource projections is the experience with North American natural gas. Natural gas supplies roughly 20 percent of U.S. energy demand. It has been plentiful at real prices of roughly $2/Mcf for almost two decades. Over the past 10 years, natural gas has become the fuel of choice for new electric power generation plants and, at present, virtually all new electric power generation plants use natural gas.

Part of the attractiveness of natural gas was resource estimates for the U.S. and Canada that promised growing supply at reasonable prices for the foreseeable future. That optimism turns out to have been misplaced, and the U.S. is now experiencing supply constraints and high natural gas prices. Supply difficulties are almost certain for at least the remainder of the decade. The North American natural gas situation provides some useful lessons relevant to the peaking of conventional world oil production.

B. The Optimism

As recently as 2001, a number of credible groups were optimistic about the ready availability of natural gas in North America. For example:

- In 1999 the National Petroleum Council stated "U.S. production is projected to increase from 19 trillion cubic feet (Tcf) in 1998 to 25 Tcf in 2010 and could approach 27 Tcf in 2015.... Imports from

Canada are projected to increase from 3 Tcf in 1998 to almost 4 Tcf in 2010." [51]

- In 2001 Cambridge Energy Research Associates (CERA) stated "The rebound in North American gas supply has begun and is expected to be maintained at least through 2005. In total, we expect a combination of US lower-48 activity, growth in Canadian supply, and growth in LNG imports to add 8.95 Bcf per day of production by 2005." [52]
- The U.S. Energy Department's Energy Information Administration (EIA) in 1999 projected that U.S. natural gas production would grow continuously from a level of 19.4 Tcf in 1998 to 27.1 Tcf in 2020.[53]

C. Today's Perspectives

The current natural gas supply outlook has changed dramatically. Among those that believe the situation has changed for the worse are the following:

- CERA now finds that "The North American natural gas market is set for the longest period of sustained high prices in its history, even adjusting for inflation. Disappointing drilling results … have caused CERA to revise the outlook for North American supply downward… The downward revisions represent additional disappointing supply news, painting a more constrained picture for continental supply. Gas production in the United States (excluding Alaska) now appears to be *in permanent decline*, and modest gains in Canadian supply will not overcome the US downturn."[54]
- Raymond James & Associates finds that "Natural gas production continues to drop despite a 20 percent increase in U.S. drilling activity since April 2003."[55] "U.S. natural gas production is heading firmly downwards…"[56]
- "Lehman now expects full-year U.S. production to decline by 4% following a 6% decline in 2003. …. Domestic production is forecast to fall to 41.0 billion cubic feet a day by 2008 from 46.8 in 2003 and 52.1 in 1998. After a sharp 12% fall in 2003, Canadian imports are seen dropping..."[57]
- The NPC now contends that "Current higher gas prices are the result of a fundamental shift in the supply and demand balance.

> North America is moving to a period in its history in which it will no longer be self-reliant in meeting its growing natural gas needs; production from traditional U.S. and Canadian basins has plateaued."[58]

Canada has been a reliable U.S. source of natural gas imports for decades. However, the Canadian situation has recently changed for the worse. For example: "Natural gas production in Alberta, the largest exporter to the huge

U.S. market, slipped 2 percent last year despite record drilling and may have peaked in 2001, the Canadian province's energy regulator said on Thursday … Production peaked at 5.1 trillion cubic feet in 2001. … (EUB) forecast flat production in 2004 and an annual decline of 2.5 percent through at least 2013."[59]

D. U.S. Natural Gas Price History

EIA data show that U.S. natural gas prices were relatively stable in constant dollars from 1987 through1998.[60] However, beginning in 2000, prices began to escalate --they were roughly 50 percent higher in 2000 compared to 1998.[61] Skipping over the recession years of 2001 and 2002, prices in late 2003 and early 2004 further increased roughly 25 percent over 2000.[62]

While it is often inappropriate to extrapolate gas or oil prices into the future based on short term experience, a number of organizations are now projecting increased U.S. natural gas prices for a number of years. For example, CERA now expects natural gas prices to rise steadily through 2007.[63]

E. LNG–Delayed Salvation

With North American natural gas production suddenly changed, hopes of meeting future demand have turned to imports of liquefied natural gas (LNG).[64] The U.S. has four operating LNG terminals, and a number of proposals for new terminals have been advanced. Indeed, the Secretary of Energy and the Chairman of the Federal Reserve Board recently called for a massive buildup in LNG imports to meet growing U.S. natural gas demand.

But the construction of new terminals demands state and local approvals. Because of NIMBYism and fear of terrorism at LNG facilities, a number of the proposed terminals have been rejected. There are also objections from Mexico, which has been proposed as a host for LNG terminals to support west coast natural gas demands.[65] In the Boston area there is an ongoing debate as to whether the nation's largest LNG terminal in Everett, Massachusetts, ought to be shut down, because of terrorist concerns.[66] Decommissioning of that terminal would exacerbate an already tight national natural gas supply situation. Public fears about LNG safety were heightened by an explosion at an LNG liquefaction plant in Algeria that killed 27 people in January 2004. Alternatively, some are considering locating LNG terminals offshore with gas pipelined underwater to land; related costs will be higher, but safety would be enhanced.

F. The U.S. Current Natural Situation

U.S. natural gas demand is increasing; North American natural gas production is declining or poised for decline as indicated in references 53, 54, and 55. The planned U.S. expansion of LNG imports is experiencing delays. U.S. natural gas supply shows every sign of deteriorating significantly before mitigation provides an adequate supply of low cost natural gas. Because of the time required to make major changes in the U.S. natural gas infrastructure and marketplace, forecasts of a decade of high prices and shortages are credible.

G. Lessons Learned

A full discussion of the complex dimensions of the current U.S. natural gas situation is beyond the scope of this study; such an effort would require careful consideration of geology, reserves estimation, natural gas exploration and production, government land restrictions, storage, weather, futures markets, etc. Nevertheless, we believe that the foregoing provides a basis for the following observations:

- Like oil reserves estimation, natural gas reserves estimation is subject to enormous uncertainty. North American natural gas reserves estimates now appear to have been excessively optimistic and North American natural gas production is now almost certainly in decline.

- High prices do not a priori lead to greater production. *Geology is ultimately the limiting factor, and geological realities are clearest after the fact.*
- Even when urgent, nation-scale energy problems arise, business-as-usual mitigation activities can be dramatically delayed or stopped by state and local opposition and other factors.

If experts were so wrong on their assessment of North American natural gas, are we really comfortable risking that the optimists are correct on world conventional oil production, which involves similar geological and technological issues?

If higher prices did not bring forth vast new supplies of North American natural gas, are we really comfortable that higher oil prices will bring forth huge new oil reserves and production, when similar geology and technologies are involved?

Chapter IV

MITIGATION OPTIONS AND ISSUES

A. CONSERVATION

Practical mitigation of the problems associated with world oil peaking must include fuel efficiency technologies that could impact on a large scale. Technologies that may offer significant fuel efficiency improvements fall into two categories: *retrofits,* which could improve the efficiency of existing equipment, and *displacement technologies*, which could replace existing, less efficient oil-consuming equipment. A comprehensive discussion of this subject is beyond the scope of this study, so we focus on what we believe to be the highest impact, existing technologies. Clearly, other technologies might contribute on a lesser scale.

From our prior discussion of current liquid fuel usage (Chapter III), it is clear that automobiles and light trucks (light duty vehicles or LDVs) represent the largest targets for consumption reduction. This should not be surprising: Auto and LDV fuel use is large, and fuel efficiency has not been a consumer priority for decades, largely due to the historically low cost of gasoline. An established but relatively little-used engine technology for LDVs in the U.S. is the diesel engine, which is up to 30 percent more efficient than comparable gasoline engines. Future U.S. use of diesels in LDVs has been problematic due to increasingly more stringent U.S. air emission requirements. European regulations are not as restrictive, so Europe has a high population of diesel LDVs – between 55 and 70 percent in some countries. [67]

A new technology in early commercial deployment is the hybrid system, based on either gasoline or diesel engines and batteries. In all-around driving tests, gasoline hybrids have been found to be 40 percent more efficient in small cars and 80 percent more efficient in family sedans.[68]

For retrofit application, neither diesel nor hybrid engines appear to have significant potential, so their use will likely be limited to new vehicles. Under business-as-usual market conditions, hybrids might reach roughly 10 percent on-the-road U.S. market share by 2015.[69] That penetration rate is based on the fact that the technology has met many of the performance demands of a significant number of today's consumers and that gasoline hybrids use readily available fuel.

Government-mandated vehicle fuel efficiency requirements are virtually certain to be an element in the mitigation of world oil peaking. One result would almost certainly be the more rapid deployment of diesel and / or hybrid engines. Market penetration of these technologies cannot happen rapidly, because of the time and effort required for manufacturers to retool their factories for large-scale production and because of the slow turnover of existing stock. In addition, a shift from gasoline to diesel fuel would require a major refitting of refineries, which would take time.

Nation-scale retrofit of existing LDVs to provide improved fuel economy has not received much attention. One retrofit technology that might prove attractive for the existing LDV fleet is "displacement on demand" in which a number of cylinders in an engine are disabled when energy demand is low. The technology is now available on new cars, and fuel economy savings of roughly 20 percent have been claimed.[70] The feasibility and cost of such retrofits are not known, so we consider this option to be speculative.

It is difficult to project what the fuel economy benefits of hybrid or diesel LDVs might be on a national scale, because consumer preferences will likely change once the public understands the potential impacts of the peaking of world oil production. For example, the current emphasis on large vehicles and SUVs might well give way to preferences for smaller, much more fuel-efficient vehicles.

The fuel efficiency benefits that hybrids might provide for heavy-duty trucks and buses are likely smaller than for LDVs for a number of reasons, including the fact that there has long been a commercial demand for higher efficiency technologies in order to minimize fuel costs for these fleets.

Hybrids can also impact the medium duty truck fleet, which is now heavily populated with diesel engines. For example, road testing of diesel hybrids in FedEx trucks recently began, with fuel economy benefits of 33 percent claimed.[71] On the other hand, there appears to be limits to the fuel economy benefits of hybrid engines in large vehicles; for example, the fuel savings in hybrid buses might only be in the 10 percent range.[72]

On the distant horizon, innovations in aircraft design may result in large fuel economy improvements. For example, a 25 to 50 percent fuel efficiency improvement may be possible with a new, blended wing aircraft.[73] Such benefits would require the purchase of entirely new equipment, requiring a decade or more for significant market penetration. Innovations for major liquid fuel savings for trains and ships may exist but are not widely publicized.

B. Improved Oil Recovery

Management of an oil reservoir over its multi-decade life is influenced by a range of factors, including 1) actual and expected future oil prices; 2) production history, geology, and status of the reservoir; 3) cost and character of production-enhancing technologies; 4) timing of enhancements; 5) the financial condition of the operator; 6) political and environmental circumstances, 7) an operator's other investment opportunities, etc.

Improved Oil Recovery (IOR) is used to varying degrees on all oil reservoirs. IOR encompasses a variety of methods to increase oil production and to expand the volume of recoverable oil from reservoirs. Options include in-fill drilling, hydraulic fracturing, horizontal drilling, advanced reservoir characterization, enhanced oil recovery (EOR), and a myriad of other methods that can increase the flow and recovery of liquid hydrocarbons. IOR can also include many seemingly mundane efficiencies introduced in daily operations.[74]

IOR technologies are adapted on a case-by-case basis. It is not possible to estimate what IOR techniques or processes might be applied to a specific reservoir without having detailed knowledge of that reservoir. Such knowledge is rarely in the public domain for the large conventional oil reservoirs in the world; if it were, then a more accurate estimate of the timing of world oil peaking would be possible.

A particularly notable opportunity to increase production from existing oil reservoirs is the use of enhanced oil recovery technology (EOR), also known as tertiary recovery. EOR is usually initiated after primary and secondary recovery have provided most of what they can provide. Primary production is the process by which oil naturally flows to the surface because oil is under pressure underground. Secondary recovery involves the injection of water into a reservoir to force additional oil to the surface.

EOR has been practiced since the 1950s in various conventional oil reservoirs, particularly in the United States. The process that likely has the largest worldwide potential is miscible flooding wherein carbon dioxide (CO_2), nitrogen or light hydrocarbons are injected into oil reservoirs where they act as solvents to move residual oil. Of the three options, CO_2 flooding has proven to be the most frequently useful. Indeed, naturally occurring, geologically sourced CO_2 has been produced in Colorado and shipped via pipeline to west Texas and New Mexico for decades for EOR. CO_2 flooding can increase oil recovery by 7-15 percent of original oil in place (OOIP).[75] Because EOR is relatively expensive, it has not been widely deployed in the past. However, in a world dealing with peak conventional oil production and higher oil prices, it has significant potential.

Because of various cost considerations, enhanced oil recovery processes are typically not applied to a conventional oil reservoir until after oil production has peaked. Therefore, EOR is not likely to increase reservoir peak production. However, EOR can increase total recoverable conventional oil, and production from the reservoirs to which it is applied does not decline as rapidly as would otherwise be the case. This concept is notionally shown in Figure II.1.

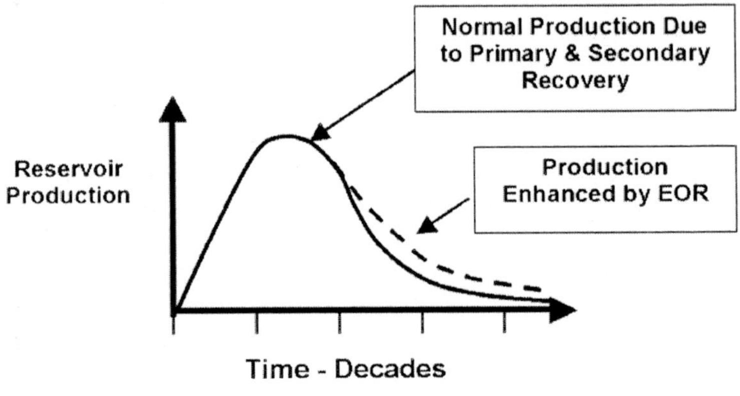

Figure IV.1. The Timing of EOR Applications

C. Heavy Oil and Oil Sands

This category of unconventional oil includes a variety of viscous oils that are called heavy oil, bitumen, oil sands, and tar sands. These oils have potential to play a much larger role in satisfying the world's needs for liquid fuels in the future.

The largest deposits of these oils exist in Canada and Venezuela, with smaller resources in Russia, Europe and the U.S. While the size of the Canadian and Venezuela resources are enormous, 3-4 trillion barrels in total, the amount of oil estimated to be economically recoverable is of the order of 600 billion barrels.[76] This relatively low fraction is in large part due to the extremely difficult task of extracting these oils.[77]

Canadian oil sands production results in a range of products, only a part of which can be refined into finished fuels that can substitute for petroleum-based fuels. These high quality oil-sands-derived products are called synthetic crude oil (SCO). Other products from oil sands processing are Dilbit, a blend of diluent and bitumen, Synbit, a blend of synthetic crude oil and bitumen, and Syndilbit, a blend of Synbit and diluent. Current Canadian production is approximately 1 million bpd of which 600,000 bpd is synthetic crude oil and 400,000 bpd is lower grade bitumen.[78]

The reasons why the production of unconventional oils has not been more extensive is as follows: 1) Production costs for unconventional oils are typically much higher than for conventional oil; 2) Significant quantities of energy are required to recover and transport unconventional oils; and 3) Unconventional oils are of lower quality and, therefore, are more expensive to refine into clean transportation fuels than conventional oils.

Canadian oil sands have been in commercial production for decades. During that time, production costs have been reduced considerably, but costs are still substantially higher than conventional oil production. Canadian oil sands production currently uses large amounts of natural gas for heating and processing. Canada recently recognized that it no longer has the large natural gas resources once thought, so oil sands producers are considering building coal or nuclear plants as substitute energy sources to replace natural gas.[79] The overall efficiency of Canadian oil sands production is not publicly available but has been estimated to be less than 70 percent for total product, only a part of which is a high-quality substitute transport fuel.[80]

In addition to needing a substitute for natural gas for processing oil sands, there are a number of other major challenges facing the expansion of Canadian oil sands production, including water [81] and diluent availability, financial capital, and environmental issues, such as SOX and NOX emissions, waste water cleanup, and brine, coke, and sulfur disposition. In addition, because Canada is a signatory to the Kyoto Protocol and because oil sands production results in significant CO2 emissions per barrel, there may be related constraints yet to be fully evaluated.

The current Canadian vision is to produce a total of about 5 MM bpd of products from oil sands by 2030. This is to include about 3 MM bpd of synthetic crude oil from which refined fuels can be produced, with the remainder being poorer quality bitumen that could be used for energy, power, and/or hydrogen and petrochemicals production. 5 MM bpd would represent a five-fold increase from current levels of production.[82] Another estimate of future production states that if all proposed oil sands projects proceed on schedule, industry could produce 3.5 MM bpd by 2017, representing 2 MM bpd of synthetic crude and 1.5 MM bpd of unprocessed lower-grade bitumen.[83] It should be noted that not everyone supports this expansion. For example, the executive director of the Sierra Club of Canada, calls tar sands "... the world's dirtiest source of oil." [84]

Venezuela's extra-heavy crude oil and bitumen deposits are situated in the Orinoco Belt, located in Central Venezuela. There are currently a number of joint ventures between the Venezuelan oil company, PdVSA, and foreign partners to develop and produce this oil. In 2003, production was about 500,000 bpd of synthetic crude oil. That is expected to increase to 600,000 bpd by 2005.[85] While the weather in tropical Venezuela is more conducive to oil production operations than the bitter winters of Alberta, Canada, the political climate in Venezuela has been particularly unsettled in recent years, which could impact future production.

In closing, it is also worth noting that the bitumen yield from oil sands surface mining operations is about 0.6 barrels per ton of mined material, excluding overburden removal. This is similar to the yield from a good quality oil shale, but is less than Fisher-Tropsch liquid yields from coal, which is about 2.6 barrels per ton of coal. [86]

D. Gas-To-Liquids (GTL)

Very large reservoirs of natural gas exist around the world, many in locations isolated from gas-consuming markets. Significant quantities of this "stranded gas" have been liquefied and transported to various markets in refrigerated, pressurized ships in the form of liquefied natural gas (LNG). Japan, followed by Korea, Spain and the U.S. were the largest importers of LNG in 2003. LNG accounted for an important fraction of all traded gas volumes in 2003, and that fraction is projected to continue to grow considerably in the future.[87]

Another method of bringing stranded natural gas to world markets is to disassociate the methane molecules, add steam, and convert the resultant mixture to high quality liquid fuels via the Fisher-Tropsch (F-T) process. As with coal liquefaction, F-T based GTL results in clean, finished fuels, ready for use in existing end-use equipment with only modest finishing and blending. This Gas-

To-Liquids process has undergone significant development over the past decade. Shell now operates a 14,500 bpd GTL plant in Malaysia. A number of large, new commercial plants recently announced include three large units in Qatar --a 140,000 bpd Shell facility, a 160,000 bpd ConocoPhillips facility, and a 120,000 bpd Marathon Oil plant. Projects under development and consideration total roughly 1.7 MM bpd, but not all will come to fruition. Under business-as-usual conditions, 1.0 MM bpd may be produced by 2015, in line with a recent estimate of 600,000 bpd of GTL diesel fuel by 2015 --the remaining 400,000 bpd being gasoline and other products.[88]

E. Liquid Fuels from U.S. Domestic Resources

The U.S. has three types of natural resource from which substitute liquid fuels can be manufactured: coal, oil shale, and biomass. All have been shown capable of producing high quality liquid fuels that can supplement or substitute for the fuels now produced from petroleum.

To derive liquid fuels from *coal*, the leading process involves gasification of the coal, removal of impurities from the resultant gas, and then synthesis of liquid fuels using the Fisher-Tropsch process. Modern gasification technologies have been dramatically improved over the years, with the result that over 150 gasifiers are in commercial operation around the world, a number operating on coal. Gas cleanup technologies are well developed and utilized in refineries worldwide. F-T synthesis is also well developed and commercially practiced. A number of coal liquefaction plants were built and operated during World War II, and the Sasol Company in South Africa subsequently built a number of larger, more modern facilities.[89] The U.S. has huge coal reserves that are now being utilized for the production of electricity; those resources could also provide feedstock for large-scale liquid fuel production.[90] Lastly, coal liquids from gasification/F-T synthesis are of such high quality that they do not need to be refined. When co-producing electricity, coal liquefaction is a developed technology, currently believed capable of providing clean substitute fuels at $30-35 per barrel.[91]

The U.S. is endowed with a vast resource of *oil shale*, located primarily in the western part of the Lower 48 states with lesser quantities in the mid

Atlantic region. Processes for mining shale and retorting it at high temperatures were developed intensively in the late 1970s and early 1980s. However, when oil prices decreased in the mid 1980s, all large-scale oil shale R&D was terminated.[92]

The oil shale processing technologies that were pursued in the past required large volumes of water, which is now increasingly scarce in the western states. Also, air emissions regulations have become much stricter in the ensuing years, presenting additional challenges for shale mining and processing. Finally, it should be noted that the oil produced from shale retorting requires refining before it can be used as transportation fuels.

In recent years, Shell has been developing a new shale oil recovery process that uses insitu heating and avoids mining and massive materials handling. Little is known about the process and its economics, so its potential cannot now be evaluated.[93] (See Appendix VI for notes on shale oil).

Biomass can be grown, collected and converted to substitute liquid fuels by a number of processes. Currently, biomass-to-ethanol is produced on a large scale to provide a gasoline additive. The market for ethanol derived from biomass is influenced by federal requirements and facilitated by generous federal and state tax subsidies. Research holds promise of more economical ethanol production from cellulosic ("woody") biomass, but related processes are far from economic. Reducing the cost of growing, harvesting, and converting biomass crops will be necessary.[94] In other parts of the world, biomass-to-liquid fuels might be more attractive, depending on a myriad of factors, including local labor costs. Related projections for large-scale production would be strictly speculative. In summary, there are no developed biomass-to-fuels technologies that are now near cost competitive. (See Appendix VI for notes on biomass).

F. Fuel Switching to Electricity

Electricity is only used to a limited extent in the transportation sector. Diesel fuels (mid-distillates) power most rail trains in the U.S.; only a modest fraction are electric powered. Other electric transportation is limited to special situations, such as forklifts, in-factory transporters, etc.

In the 1990s electric automobiles were introduced to the market, spurred by a California clean vehicle requirement. The effort was a failure because existing batteries did not provide the vehicle range and performance that customers demanded. In the future, electricity storage may improve enough to win consumer acceptance of electric automobiles. In addition, extremely high gasoline prices may cause some consumers to find electric automobiles more acceptable, especially for around-town use. Such a shift in public preferences is unpredictable, so electric vehicles cannot now be projected as a significant offset to future gasoline use.

A larger number of train routes could be outfitted for electric trains, but such a transition would likely be slow, because of the need to build additional electric power plants, transmission lines, and electric train cars. Since existing diesel locomotives use electric drive, their retrofit might be feasible. However, since diesel fuel use in trains is only roughly 0.3 MM bpd, [95] electrification of trains would not have a major impact on U.S liquid fuel consumption.

There are no known near-commercial means for electrifying heavy trucks or aircraft, so related conversions are not now foreseeable.

G. Other Fuel Switching

It is conceivable that consumers who now use mid-distillates and LPG (Liquefied Petroleum Gas) for heating could switch to natural gas or electricity, thereby freeing up liquid fuels for transportation. Analysis of this path is beyond the scope of this study, but it should be noted that these uses represent only a few percent of U.S. liquid fuel consumption. Such switching on a large scale would require the construction of compensating natural gas and/or electric power facilities and infrastructure, which would not happen quickly. In addition, freed-up liquids would likely require further refining to meet market and environmental requirements. Related refining would require refinery construction, which would also be time consuming.

H. Hydrogen

Hydrogen has potential as a long-term alternative to petroleum-based liquid fuels in some transportation applications. Like electricity, hydrogen is an energy carrier; hydrogen production requires an energy source for its production. Energy sources for hydrogen production include natural gas, coal, nuclear power, and renewables. Hydrogen can be used in internal combustion engines, similar to those in current use, or via chemical reactions in fuel cells.

The Department of Energy is currently conducting a high profile program aimed at developing a "hydrogen economy." [96] DOE's primary emphasis is on hydrogen for light duty vehicle application (automobiles and light duty trucks). Recently, the National Research Council (NRC) completed a study that included an evaluation of the technical, economic and societal challenges associated with the development of a hydrogen economy.[97] That study is the basis for the following highlights.

A lynchpin of the current DOE hydrogen program is fuel cells. In order for fuel cells to compete with existing petroleum-based internal combustion engines, particularly for light duty vehicles, the NRC concluded that fuel cells must improve by 1) a factor of 10-20 in cost, 2) a factor of five in lifetime, and 3) roughly a factor of two in efficiency. The NRC did not believe that such improvements could be achieved by technology development alone; instead, new concepts (breakthroughs) will be required. In other words, today's technologies do not appear practically viable.[98]

Because of the need for unpredictable inventions in fuel cells, as well as viable means for on-board hydrogen storage, the introduction of commercial hydrogen vehicles cannot be predicted.

I. Factors That Can Cause Delay

It is extremely difficult, expensive, and time consuming to construct any type of major energy-related facility in the U.S. today. Even assuming the expenditure of substantial time and money, it is not certain that many proposed facilities will ever be constructed. The construction of transmission lines, interim and permanent nuclear waste disposal facilities, electric generation plants, waste incinerators, oil refineries, LNG terminals, waste recycling facilities, petrochemical plants, etc. is increasingly problematic.

What used to be termed the "not-in-my-back-yard" (NIMBY) principle has evolved into the "build-absolutely-nothing-anywhere-near-anything" (BANANA) principle, which is increasingly being applied to facilities of any type, including low-income housing, cellular phone towers, prisons, sports stadiums, water treatment facilities, airports, hazardous waste facilities, and even new fire houses.[99] Construction of even a single, relatively innocuous, urgently needed facility can easily take more than a decade. For example, in 1999, King County,

Washington, initiated the siting process for the Brightwater wastewater treatment plant, which it hopes to have operation in 2010.[100]

The routine processes required for siting energy facilities can be daunting, expensive, and time consuming, and if a facility is at all controversial, which is almost invariably the case, opponents can often extend the permitting process until sponsors terminate their plans. For example, approval for new, small, distributed energy systems requires a minimum of 18 separate steps, requiring approval from four federal agencies, 11 state government agencies, and 14 local government agencies.[101] Opponents of energy facilities routinely exercise their right to raise objections and offer alternatives. Intervenors in permitting processes may delay decisions and in some cases force outright cancellations, although cases do exist in which facilities have been sited quickly.

The implications for U.S. homeland-based mitigation of world oil peaking are troubling. To replace dwindling supplies of conventional oil, large numbers of expensive and environmentally intrusive substitute fuel production facilities will be required. Under current conditions, it could easily require more than a decade to construct a large coal liquefaction plant in the U.S. The prospects for constructing 25-50, with the first ones coming into operation within a three year time window are essentially nil. *Absent change, the U.S. may end up on the path of least resistance, allowing only a few substitute fuels plants to be built on U.S. soil; in the process the U.S. would be adding substitute fuel imports to its increasing dependence on imports of conventional oil.*

For the U.S. to attain a lower level of dependence on liquid fuel imports after the advent of world oil peaking, a major paradigm shift will be required in the current approach to the construction of capital-intensive energy facilities. Federal and state governments will have to adopt legislation allowing the acceleration of the development of substitute fuels projects from current decade time-scales. During World War II, facilities of all types were constructed on a scale and schedules that would have previously been inconceivable. In the face of the 1973 energy crisis, the Alaska oil pipeline was approved and constructed in record time.[102]

While world oil peaking poses many dangers for the U.S., it also offers substantial opportunities. The U.S. could emerge as the world's largest producer of substitute liquid fuels, if it were to undertake a massive program to construct substitute fuel production facilities on a timely basis. The nation is ideally positioned to do so because it has the world's largest coal reserves, and it could muster the required capital, technology, and labor to implement such a program. However, unless a process is developed to expedite plant construction, this opportunity could easily slip away. Other nations, such as China, India, Japan, Korea, and others also have the capabilities needed to construct and operate such plants. Under current conditions, other countries are able to bring such large energy projects on-line much more rapidly than the U.S. Such countries could conceivably even import U.S. coal, convert it to liquid fuels products, and then export finished product back to the U.S. and elsewhere.

The U.S. has well-developed coal mining, transportation, and shipping systems that move coal to the highest bidders, be they domestic or international. As recently as 1981, 14 percent of U.S. coal production was exported.[103] While that number has declined in recent years, the U.S. could easily expand its current coal exports many fold to provide feedstock for coal liquefaction plants in other nations. Not only would the U.S. be dependent on foreign sources for conventional oil, which will continue to dwindle in volume after peaking, but it could also become dependant on foreign sources for substitute fuels derived from U.S. coal.

Chapter V

A World Problem

Oil is essential to all countries. In 2002 daily consumption ranged from almost 20 million barrels in the U.S. to 20 barrels in the tiny South Pacific island of Niue, population 2,400.[104]

Table V.1.Top World Oil Producing and Consuming Countries -2002

Producers				Consumers			
Rank	Country	MM bpd	Percent	Rank	Country	MM bpd	Percent
1	United States	9.0	11.7	1	United States	19.8	25.3
2	Saudi Arabia	8.7	11.3	2	Japan	5.3	6.8
3	Russia	7.7	10.0	3	China	5.2	6.6
4	Mexico	3.6	4.7	4	Germany	2.7	3.5
5	Iran	3.5	4.6	5	Russia	2.6	3.3
6	China	3.5	4.6	6	India	2.2	2.8
7	Norway	3.3	4.3	7	Korea, South	2.2	2.8
8	Canada	2.9	3.8	8	Brazil	2.2	2.8
9	Venezuela	2.9	3.8	9	Canada	2.1	2.7
10	United Kingdom	2.6	3.3	10	France	2.0	2.5
11	United Arab Emirates	2.4	3.1	11	Mexico	2.0	2.5
12	Nigeria	2.1	2.8	12	Italy	1.8	2.4
13	Iraq	2.0	2.7	13	United Kingdom	1.7	2.2
14	Kuwait	2.0	2.6	14	Saudi Arabia	1.5	1.9
15	Brazil	1.8	2.3	15	Spain	1.5	1.9
16	Algeria	1.6	2.0	16	Iran	1.3	1.7
17	Libya	1.4	1.8	17	Indonesia	1.1	1.4
18	Indonesia	1.4	1.8	18	Taiwan	0.9	1.2
19	Kazakhstan	0.9	1.2	19	Netherlands	0.9	1.1
20	Oman	0.9	1.2	20	Australia	0.9	1.1
	103 other countries	***12.6***	***16.3***		***194 other countries***	***18.4***	***23.5***

Oil is produced in 123 countries. The top 20 producing countries provide over 83 percent of total world oil. Production by the largest producers is shown in Table V.1.[105] The table also lists the top 20 oil-consuming countries and their respective consumption. In total, the top 20 countries consume over 75 percent of the average daily production. Beyond these larger consumers, oil is also utilized in all the world's 194 remaining countries.

Chapter VI

THREE SCENARIOS

A. INTRODUCTION

Issues related to the peaking of world oil production are extremely complex, involve literally trillions of dollars and are very time-dependent. To explore these matters, we selected three mitigation scenarios for analysis:

- Scenario I assumes that action is not initiated until peaking occurs.
- Scenario II assumes that action is initiated 10 years before peaking.
- Scenario III assumes action is initiated 20 years before peaking.

Our approach is simplified in order to provide transparency and promote understanding. Our estimates are approximate, but the mitigation envelope that results is believed to be indicative of the realities of such an enormous undertaking.

B. MITIGATION OPTIONS

Our focus is on large-scale, physical mitigation, as opposed to policy actions, e.g. tax credits, rationing, automobile speed restrictions, etc. We define *physical mitigation* as 1) implementation of technologies that can substantially reduce the consumption of liquid fuels (improved fuel efficiency) while still delivering comparable service and 2) the construction and operation of facilities that yield large quantities of liquid fuels.

C. Mitigation Phase-In

The pace that governments and industry chose to mitigate the negative impacts of the peaking of world oil production is to be determined.. As a limiting case, we choose *overnight go-ahead decision-making* for all actions, i.e., crash programs. Our rationale is that in a sudden disaster situation, crash programs are most likely to be quickly implemented. Overnight go-ahead decision-making is most probable in our Scenario I, which assumes no action prior to the onset of peaking. By assuming overnight implementation in all three of our scenarios, we avoid the arduous and potentially arbitrary challenge of developing a more likely, real world decision-making sequence. This is obviously an *optimistic assumption* because government and corporate decision-making is never instantaneous.

D. The Use of Wedges

The model chosen to illustrate the possible effects of likely mitigation actions involves the use of "delayed wedges" to approximate the scale and pace of each action. The use of wedges was effectively utilized in a recent paper by Pacala and Socolow.[106]

Our wedges are composed of two parts. The first is the preparation time needed prior to tangible market penetration. In the case of efficient transportation, this time is required to redesign vehicles and retool factories to produce more efficient vehicles. In the case of the production of substitute fuels, the delay is associated with planning and construction of relevant facilities.

After the preparation phase, our wedges then approximate the penetration of mitigation effects into the marketplace. This might be the growing sales of more fuel-efficient vehicles or the growing production of substitute fuels. Our wedge pattern is shown in Figure VI.1., where the horizontal axis is time and the vertical axis is market impact, measured in barrels per day of savings or production. The figure is bounded on the right side for illustrative purposes only. We assume our wedges continue to expand for a few decades, which simplifies illustration but is increasingly less realistic over time because markets will adjust and impact rates will change.

How our delayed wedges approximate reality is illustrated in Figure VI.2, which shows possible fuel savings associated with implementation of significant new Corporate Average Fuel Efficiency (CAFE) standards.[107]

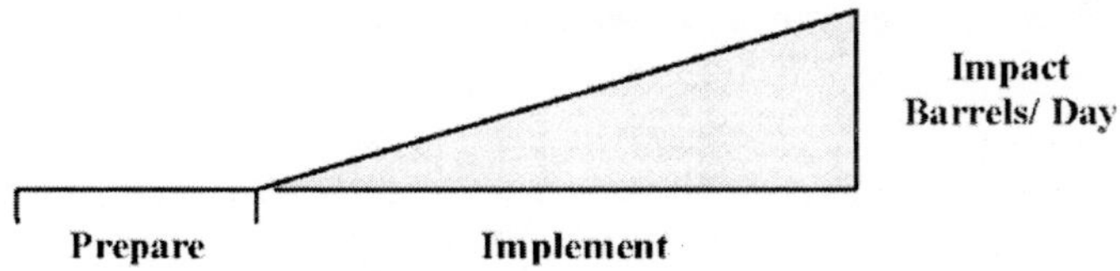

Figure VI.1. Delayed wedge approximation for various mitigation options

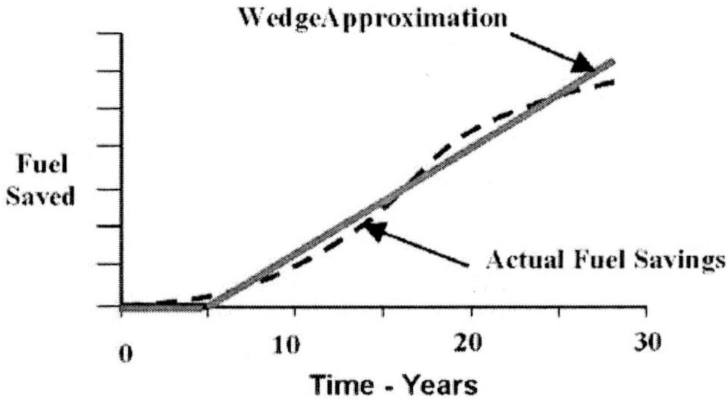

Figure VI.2. The delayed wedge approximation in the case of major changes in transportation fuel consumption

Our aim is to approximate reality in a simple manner. Greater detail is beyond the scope of this study and would require in-depth analysis.

E. Criteria for Wedge Selection

Our criteria for selecting candidates for our energy saving and substitute oil production wedges were as follows:

1. The option must produce liquid fuels that can, as produced or as refined, substitute for liquid fuels currently in widespread use, e.g. gasoline, jet fuel, diesel, etc. The end products will thus be compatible with existing distribution systems and end-use equipment.
2. The option must be capable of liquid fuels savings or production on a massive scale – ultimately millions to tens of millions of barrels per day worldwide.
3. The option must include technology that is commercial or near commercial, which at a minimum requires that the process has been

demonstrated at commercial scale. For production technologies, this means that at least one plant has operated at greater than 10,000 bpd for at least two years, and product prices from the process are less than $50/barrel in 2004 dollars. For fuels efficiency technologies, the technology must have at least entered the commercial market by 2004.

4. Substitute fuel production technologies must be inherently energy efficient, which we assume to mean that greater than 50 percent of process energy input is contained in the clean liquid fuels product.[108]
5. The option must be environmentally clean by 2004 standards.
6. While domestic resources are of greatest interest to the U.S., the oil market is international, so substitute fuel feedstocks not abundantly available in the U.S. must also be considered, e.g. heavy oil/tar sands and gas-to-liquids.
7. Energy sources or energy efficiency technologies that produce or save electricity are not of interest in this context because commercial processes to convert electricity to clean hydrocarbon fuels do not currently exist.

F. Wedges Selected & Rejected

The combination of technologies, processes, and feedstocks that meet these criteria are as follows:

1. Fuel efficient transportation,
2. Heavy oil/Oil sands,
3. Coal liquefaction,
4. Enhanced oil recovery,
5. Gas-to-liquids.

In the end-use category, a dramatic increase in the efficiency of petroleum-based fuel equipment is one attractive option. As previously described, the imposition of CAFE requirements for automobile in 1975 was one of the most effective of the government mandates initiated in response to the 1973-74 oil embargo. In recent years, fuel economy for automobiles has not been a high national priority in the U.S. Nevertheless, a new hybrid engine technology has been phasing into the automobile and truck markets. In a period of national oil emergency, hybrid technology could be massively implemented for new vehicle applications. Hybrid technologies offer fuel economy improvements of 40 percent or more for automobiles and light-medium trucks – no other engine technologies offer such large, near-term fuel economy benefits.[109]

The fuels production options that we chose are heavy oil/tar sands, coal liquefaction, improved oil recovery, and gas-to-liquids. Our rationale was as follows:

1. Enhanced Oil Recovery is applicable worldwide.
2. Heavy oil / Oil sands is currently commercial in Canada and Venezuela.
3. Coal liquefaction is a well-developed, near-commercial technology.
4. Gas-To-Liquids is commercially applicable where natural gas is remote from markets.

We excluded a number of options for various reasons. While the U.S. has a huge resource of shale oil that could be processed into substitute liquid fuels, the technology to accomplish that task is not now ready for deployment. Because various shale oil processing prototypes were developed in years past and because shale oil processing is likely to be economically attractive, a concerted effort to develop shale oil technology could well lead to shale oil becoming a contributor in Scenarios II or III. However, that would require the initiation of a major R & D program in the near future.

Biomass options capable of producing liquid fuels were also excluded. Ethanol from biomass is currently utilized in the transportation market, not because it is commercially competitive, but because it is mandated and highly subsidized. Biodiesel fuel is a subject of considerable current interest but it too is not yet commercially viable. Again, a major R & D effort might change the biomass outlook, if initiated in the near future.[110]

Over 45% of world oil consumption is for non-transportation uses. Fuel switching away from non-transportation uses of liquid fuels is likely to occur, mimicking shifts that have already taken place in the U.S. The time frame for such shifts is uncertain. For significant world scale impact, alternate large energy facilities would have to be constructed to provide the substitute energy, and that facility construction would require the kind of decade-scale time periods required for oil peaking mitigation.

Nuclear power, wind and photovoltaics produce electric power, which is not a near-term substitute fuel in transportation equipment that requires liquid fuels. In the many-decade future after oil peaking, it is conceivable that a massive shift from liquid fuels to electricity might occur in some applications. However, consideration of such changes would be speculative at this time.

It is possible that technology innovations resulting from aggressive future research may well change the outlook for various technologies in the future. *Our focus on the currently viable is in no way intended to prejudice other future options* We have chosen not to add a wedge for undefined technologies that might result from accelerated research, because such a wedge would be purely speculative. No matter what the new technology(s), implementation delay times and contribution growth rates will inherently be of the same order of magnitude of the technologies that we have considered, because of the inherent scale of all physical mitigation.

G. Modeling World Oil Supply/Demand

It is not possible to predict with certainty when world conventional oil peaking will occur or how rapidly production will decline after the peak. To develop our scenarios, *we utilize the U.S. Lower 48 production pattern as a surrogate for the world.* This assumption is justified on the basis that Lower 48 oil production represents what really happened in a large, complex oil province over the course of decades of modern oil production development.

Our starting point is the triangular pattern of production increase followed by production decline shown in Figure I.2. Our horizontal axis is centered on the year of peaking (the date is not specified) and spans plus and minus two decades. For this study, our vertical axis is pegged at a peak world oil production of 100 MM bpd, which is 18 MM bpd above the current 82 MM bpd world production. If peaking were to occur soon, 100 MM bpd might be high by 20 percent. If peaking were to occur at 125 MM bpd at some future date, the 100 MM bpd assumption would be low by 20 percent. Since the estimates in our wedges are rough under any conditions, a 100 MM bpd peak represents a credible assumption for this kind of analysis. The selection of 100 MM bpd is not intended as a prediction of magnitude or timing; its use is for illustration purposes only.

Next is the important issue of the slopes of the production profile showing the rate of growth of production/demand before peaking and the subsequent decline in production. The World Energy Council stated: "Oil demand is projected to increase at about 1.9 percent per year rising from about 75.7 million b/d in 2000 (actual) to 113-115 million b/d in 2020 – an increase of about 37.5-39.5 million b/d." [111] Recent trends indicate a 3+ percent world oil demand growth, driven in part by rapidly increasing oil consumption in China and India. However, a 3+ percent growth rate on a continuing basis seems excessive. On this basis, we assume a two percent demand growth before peaking, and we assume an intrinsic two percent long-run hypothetical, healthy economy demand after peaking. This extrapolation of demand after peaking provides a reference that facilitates calculation of supply shortfalls. The assumption has the benefit of simplicity, but it ignores the real-world feedback of oil price escalation on demand, which is sure to happen but the calculation thereof will be complicated and was beyond the scope of this study.

Estimating a decline rate after world oil production peaking is a difficult issue. While human activity dominates the demand for oil, the "rocks" (geology) will dominate the decline of world conventional oil production after peaking. Referring to U.S. Lower 48 production history, the decline after the 1970 peaking was roughly 1.7 percent per year, which we have chosen to round off to two percent per year as our estimated world conventional oil decline rate.[112] It should be noted that other analysts have projected decline rates of 3-8%, which would make the mitigation problem much more difficult.[113]

H. Our Wedges

In Appendix IV we develop the sizes of the wedges that we believe appropriate for our trends analysis. The categories, delays and 10-year estimated impacts are shown in Figure VI.3. Once again, bear in mind that these are rough approximations aimed at illustrating the inherently large scale of mitigation.

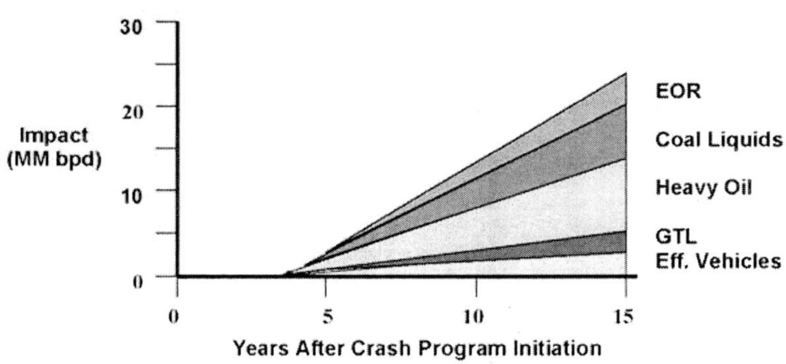

Figure VI.3. Assumed wedges

I. The Three Scenarios

As noted, our three scenarios are benchmarked to the unknown date of peaking:

- Scenario I: Mitigation begins at the time of peaking;
- Scenario II: Mitigation starts 10 years before peaking;
- Scenario III: Mitigation starts 20 years before peaking.

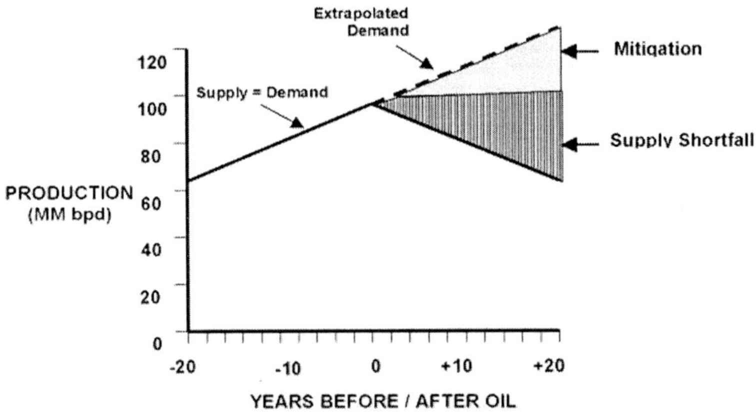

Figure VI.4. Mitigation crash programs started at the time of world oil peaking: A significant supply shortfall occurs over the forecast period.

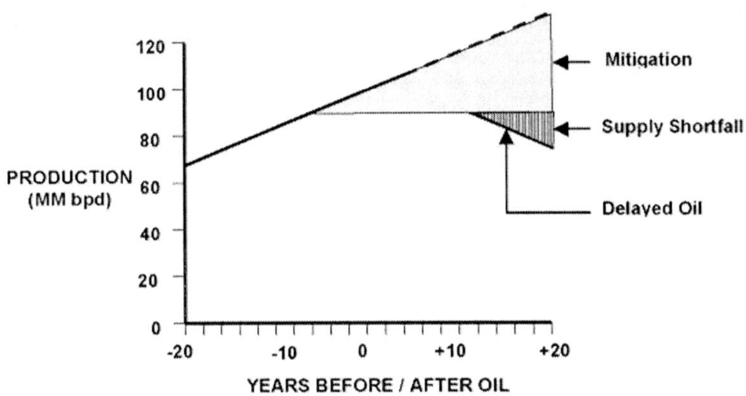

Figure VI.5. Mitigation crash programs started 10 years before world oil peaking: A moderate supply shortfall occurs after roughly 10 years.

Our mitigation choices then map onto our assumed world oil peaking pattern as shown in Figures VI.4, 5 and 6.

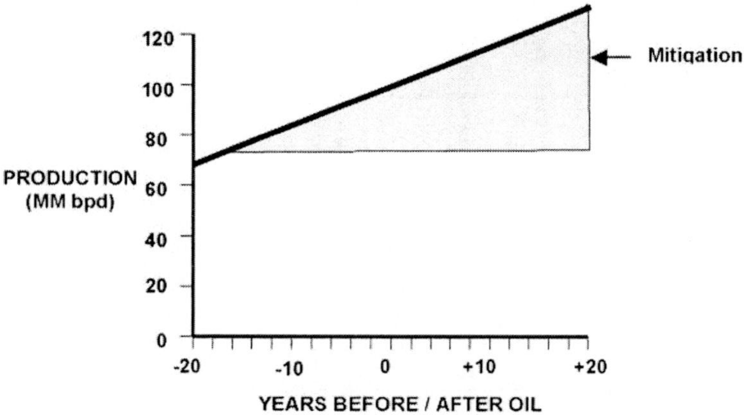

Figure VI.6. Mitigation crash programs started 20 years before world oil peaking: No supply shortfall occurs during the forecast period.

J. Observations & Conclusions on Scenarios

This exercise was conducted bottom–up; we estimated reasonable potential contributions from each viable option, summed them, and then applied them to our assumed world oil peaking pattern.

While our option contribution estimates are clearly approximate, in total they probably represent a realistic portrayal of what might be achieved with an array of physical mitigation options. Together, implementation of all of the specified options would provide 15 – 20 MM bpd impact, ten years after simultaneous initiation. Roughly 90 percent would result from substitute liquid fuel production and roughly ten percent would come from transportation fuel efficiency improvements.

Our results are congruent with the fundamentals of the problem:

- *Waiting until world oil production peaks before taking crash program action leaves the world with a significant liquid fuel deficit for more than two decades.*

- *Initiating a mitigation crash program 10 years before world oil peaking helps considerably but still leaves a liquid fuels shortfall roughly a decade after the time that oil would have peaked.*
- *Initiating a mitigation crash program 20 years before peaking appears to offer the possibility of avoiding a world liquid fuels shortfall for the forecast period.*

The obvious conclusion from this analysis is that with adequate, timely mitigation, the costs of peaking can be minimized. If mitigation were to be too little, too late, world supply/demand balance will be achieved through massive demand destruction (shortages), which would translate to significant economic hardship, as discussed earlier.

K. Risk Management

It is possible that peaking may not occur for several decades, but it is also possible that peaking may occur in the near future. We are thus faced with a daunting risk management problem:

- On the one hand, mitigation initiated soon would be premature if peaking is still several decades away.
- On the other hand, if peaking is imminent, failure to initiate mitigation quickly will have significant economic and social costs to the U.S. and the world.

The two risks are asymmetric:

- *Mitigation actions initiated prematurely will be costly and could result in a poor use of resources.*
- *Late initiation of mitigation may result in severe consequences.*

The world has never confronted a problem like this, and the failure to act on a timely basis could have debilitating impacts on the world economy. Risk minimization requires the implementation of mitigation measures well prior to peaking. Since it is uncertain when peaking will occur, the challenge is indeed significant.

Chapter VII

MARKET SIGNALS AS PEAKING IS APPROACHED

As world oil peaking is approached and demand for conventional oil begins to exceed supply, oil prices will rise steeply. As discussed in Chapter IV, related price increases are almost certain to have negative impacts on the U.S. and world economies. Another likely signal is substantially increased oil price volatility.

Oil prices have traditionally been volatile. Causes include political events, weather, labor strikes, infrastructure problems, and fears of terrorism.[114] In an era where supply was adequate to meet demand and where there was excess production capacity in OPEC, those effects were relatively short-lived. However, as world oil peaking is approached, excess production capacity by definition will disappear, so that even minor supply disruptions will cause increased price volatility as traders, speculators, and other market participants react to supply/demand events. Simultaneously, oil storage inventories are likely to decrease, further eroding security of supply, aggravating price volatility, and further stimulating speculation.[115]

While it is recognized that high oil prices will have adverse effects, the effects of increased price volatility may not be sufficiently appreciated. Higher oil price volatility can lead to reduction in investment in other parts of the economy, leading in turn to a long-term reduction in supply of various goods, higher prices, and further reduced macroeconomic activity. Increasing volatility has the potential to increase both economic disruption and transaction costs for both consumers and producers, adding to inflation and reducing economic growth rates.[116]

The most relevant experience was during the 1970s and early 1980s, when oil prices increased roughly six-fold and oil price volatility was aggravated. Those reactions have often been dismissed as a "panic response," but that experience may nevertheless be a good indicator of the oil price volatility to be expected when demand exceeds supply after oil peaking.[117]

The factors that cause oil price escalation and volatility could be further exacerbated by terrorism. For example, in the summer of 2004, it was estimated that the threat of terrorism had added a premium of 25 -33 percent to the price of a barrel of oil.[118] As world oil peaking is approached, it is not difficult to imagine that the terrorism premium could increase even more.

In conclusion, oil peaking will not only lead to higher oil prices but also to increased oil price volatility. In the process, oil could become the price setter in the broader energy market, in which case other energy prices could well become increasingly volatile and unpredictable.[119]

Chapter VIII

WILD CARDS

There are a number of factors that could conceivably impact the peaking of world oil production. Here is a list of possible upsides and downsides.

A. UPSIDES–THINGS THAT MIGHT EASE THE PROBLEM OF WORLD OIL PEAKING

The pessimists are wrong again and peaking does not occur for many decades.

- Middle East oil reserves are much higher than publicly stated.
- A number of new super-giant oil fields are found and brought into production, well before oil peaking might otherwise have occurred.
- High world oil prices over a sustained period (a decade or more) induce a higher level of structural conservation and energy efficiency.
- The U.S. and other nations decide to institute significantly more stringent fuel efficiency standards well before world oil peaking.
- World economic and population growth slows and future demand is much less than anticipated.
- China and India decide to institute vehicle efficiency standards and other energy efficiency requirements, reducing the rate of growth of their oil requirements.

- Oil prices stay at a high enough level on a sustained basis so that industry begins construction of substitute fuels plants well before oil peaking.
- Huge new reserves of natural gas are discovered, a portion of which is converted to liquid fuels.
- Some kind of scientific breakthrough comes into commercial use, mitigating oil demand well before oil production peaks.

B. Downsides-Things That Might Exacerbate the Problem of World Oil Peaking

- World oil production peaking is occurring now or will happen soon.
- Middle East reserves are much less than stated.
- Terrorism stays at current levels or increases and concentrates on damaging oil production, transportation, refining and distribution.
- Political instability in major oil producing countries results in unexpected, sustained world-scale oil shortages.
- Market signals and terrorism delay the realization of peaking, delaying the initiation of mitigation.
- Large-scale, sustained Middle East political instability hinders oil production.
- Consumers demand even larger, less fuel-efficient cars and SUVs.
- Expansion of energy production is hindered by increasing environmental challenges, creating shortages beyond just liquid fuels.

Chapter IX

SUMMARY AND CONCLUDING REMARKS

Our analysis leads to the following conclusions and final thoughts.

1. WORLD OIL PEAKING IS GOING TO HAPPEN

World production of conventional oil will reach a maximum and decline thereafter. That maximum is called the peak. A number of competent forecasters project peaking within a decade; others contend it will occur later. Prediction of the peaking is extremely difficult because of geological complexities, measurement problems, pricing variations, demand elasticity, and political influences. Peaking will happen, but the timing is uncertain.

2. OIL PEAKING COULD COST THE U.S. ECONOMY DEARLY

Over the past century the development of the U.S. economy and lifestyle has been fundamentally shaped by the availability of abundant, low-cost oil. Oil scarcity and several-fold oil price increases due to world oil production peaking could have dramatic impacts. The decade after the onset of world oil peaking may resemble the period after the 1973-74 oil embargo, and the economic loss to the United States could be measured on a trillion-dollar scale. Aggressive, appropriately timed fuel efficiency and substitute fuel production could provide substantial mitigation.

3. Oil Peaking Presents a Unique Challenge

The world has never faced a problem like this. Without massive mitigation more than a decade before the fact, the problem will be pervasive and will not be temporary. Previous energy transitions (wood to coal and coal to oil) were gradual and evolutionary; oil peaking will be abrupt and revolutionary.

4. The Problem is Liquid Fuels

Under business-as-usual conditions, world oil demand will continue to grow, increasing approximately two percent per year for the next few decades. This growth will be driven primarily by the transportation sector. The economic and physical lifetimes of existing transportation equipment are measured on decade time-scales. Since turnover rates are low, rapid changeover in transportation end-use equipment is inherently impossible.

Oil peaking represents a liquid fuels problem, not an "energy crisis" in the sense that term has been used. Motor vehicles, aircraft, trains, and ships simply have no ready alternative to liquid fuels. Non-hydrocarbon-based energy sources, such as solar, wind, photovoltaics, nuclear power, geothermal, fusion, etc. produce electricity, not liquid fuels, so their widespread use in transportation is at best decades away. Accordingly, mitigation of declining world oil production must be narrowly focused.

5. Mitigation Efforts Will Require Substantial Time

Mitigation will require an intense effort over decades. This inescapable conclusion is based on the time required to replace vast numbers of liquid fuel consuming vehicles and the time required to build a substantial number of substitute fuel production facilities. Our scenarios analysis shows:

- Waiting until world oil production peaks before taking crash program action would leave the world with a significant liquid fuel deficit for more than two decades.

- Initiating a mitigation crash program 10 years before world oil peaking helps considerably but still leaves a liquid fuels shortfall roughly a decade after the time that oil would have peaked.
- Initiating a mitigation crash program 20 years before peaking appears to offer the possibility of avoiding a world liquid fuels shortfall for the forecast period.

The obvious conclusion from this analysis is that with adequate, timely mitigation, the economic costs to the world can be minimized. If mitigation were to be too little, too late, world supply/demand balance will be achieved through massive demand destruction (shortages), which would translate to significant economic hardship.

There will be no quick fixes. Even crash programs will require more than a decade to yield substantial relief.

6. Both Supply and Demand Will Require Attention

Sustained high oil prices will stimulate some level of forced demand reduction. Stricter end-use efficiency requirements can further reduce embedded demand, but substantial, world-scale change will require a decade or more. Production of large amounts of substitute liquid fuels can and must be provided. A number of commercial or near-commercial substitute fuel production technologies are currently available, so the production of large amounts of substitute liquid fuels is technically and economically feasible, albeit time-consuming and expensive.

7. It Is a Matter of Risk Management

The peaking of world conventional oil production presents a classic risk management problem:

- Mitigation efforts initiated earlier than required may turn out to be premature, if peaking is long delayed.
- On the other hand, if peaking is imminent, failure to initiate timely mitigation could be extremely damaging.

Prudent risk management requires the planning and implementation of mitigation well before peaking. Early mitigation will almost certainly be less expensive and less damaging to the world's economies than delayed mitigation.

8. GOVERNMENT INTERVENTION WILL BE REQUIRED

Intervention by governments will be required, because the economic and social implications of oil peaking would otherwise be chaotic. The experiences of the 1970s and 1980s offer important lessons and guidance as to government actions that might be more or less desirable. But the process will not be easy. Expediency may require major changes to existing administrative and regulatory procedures such as lengthy environmental reviews and lengthy public involvement.

9. ECONOMIC UPHEAVAL IS NOT INEVITABLE

Without mitigation, the peaking of world oil production will almost certainly cause major economic upheaval. However, given enough lead-time, the problems are soluble with existing technologies. New technologies are certain to help but on a longer time scale. Appropriately executed risk management could dramatically minimize the damages that might otherwise occur.

10. MORE INFORMATION IS NEEDED

The most effective action to combat the peaking of world oil production requires better understanding of a number of issues. Is it possible to have relatively clear signals as to when peaking might occur? It would be desirable to have potential mitigation actions better defined with respect to cost, potential capacity, timing, etc. Various risks and possible benefits of possible mitigation actions need to be examined. (See Appendix V for a list of possible follow-on studies).

The purpose of this analysis was to identify the critical issues surrounding the occurrence and mitigation of world oil production peaking. We simplified many of the complexities in an effort to provide a transparent analysis. Nevertheless, our study is neither simple nor brief. We recognize that when oil prices escalate dramatically, there will be demand and economic impacts that will alter our simplified analysis. Consideration of those feedbacks will be a daunting task but one that should be undertaken.

Our study required that we make a number of assumptions and estimates. We well recognize that in-depth analyses may yield different numbers. Nevertheless, this analysis clearly demonstrates that the key to mitigation of world oil production peaking will be the construction a large number of substitute fuel production facilities, coupled to significant increases in transportation fuel efficiency. The time required to mitigate world oil production peaking is measured on a decade time-scale, and related production facility size is large and capital intensive. How and when governments decide to address these challenges is yet to be determined.

Our focus on existing commercial and near-commercial mitigation technologies illustrates that a number of technologies are currently ready for immediate and extensive implementation. Our analysis was not meant to be limiting. We believe that future research will provide additional mitigation options, some possibly superior to those we considered. Indeed, it would be appropriate to greatly accelerate public and private oil peaking mitigation research. However, the reader must recognize that doing the research required to bring new technologies to commercial readiness takes time under the best of circumstances. Thereafter, more than a decade of intense implementation will be required for world scale impact, because of the inherently large scale of world oil consumption.

APPENDICES

APPENDIX I
MOST MEANINGFUL EIA OIL PEAKING CASE

In the year 2000, EIA developed 12 scenarios for world oil production peaking using three U.S. Geological Survey (USGS) estimates of the world conventional oil resource base (Low, Expected, and High) and four annual world oil demand growth rates (0, 1, 2, and 3 percent per year).[120] We believe the most likely of the EIA scenarios is the one based on the USGS *expected ultimate world recoverable oil of 3,003 billion barrels* coupled with a *2% annual world oil demand escalation.*

Figure A-I shows the two EIA scenarios based on these assumptions. The difference between the two profiles is attributable to two assumed production decay rates following peak production. Both curves assume a 2 percent per year growth from the year 2000 until the peak. One scenario assumes a 2 percent decline after the world oil production peak, while the other assumes a steeper drop after the world oil production peak. Because the areas under both curves must equal the projected 3,003 billion barrels of recoverable conventional oil from the year 2000 forward, the rapid decay curve will inherently yield the later occurring, higher world oil production peak.

The EIA scenario that peaks in 2016 looks like the relatively symmetric U.S. Lower 48 production profile in Figure I.2. The EIA scenario that peaks in 2037 not only differs dramatically from the U.S. experience, it differs from typical individual oil reservoir experience, which often displays a relatively symmetric production profile, not the sharp drop illustrated in the alternate EIA case. On this basis, we believe that the EIA 2016 peaking case appears much more credible than the 2037 peaking case. The associated 21-

year difference between the two predicted production peaks clearly would have profound implications for the time available for mitigation.

It is worth noting that the USGS mean estimate for the remaining recoverable world oil resource is much higher than estimates made by other investigators, according to K.S. Deffeyes, retired Shell geologist and emeritus Princeton geology professor.[121] Deffeyes also opined "... in 2000 the USGS again released implausibly large estimates of world oil." A lower total reserves estimate would of course mean a world oil production peak earlier than 2016.

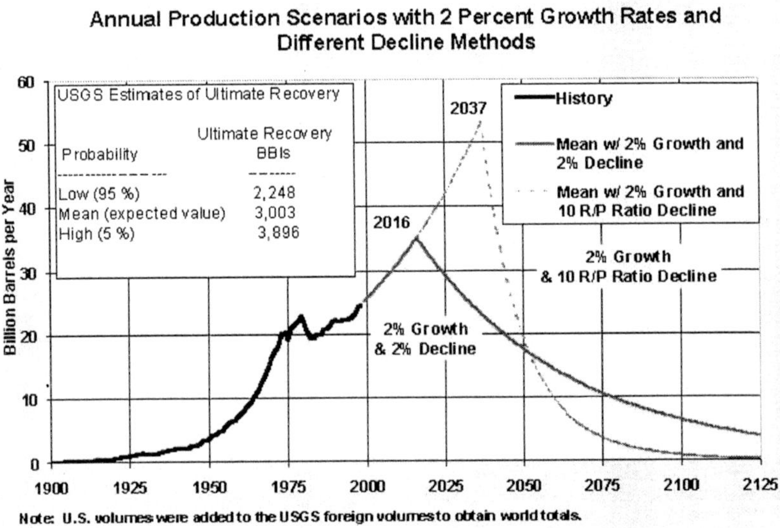

Figure A-I. Two EIA oil production scenarios based on expected ultimate world-recoverable oil of 3,003 billion barrels and a 2 percent annual world oil demand escalation

APPENDIX II
MORE HISTORICAL OIL CRISIS CONSIDERTIONS

Economists have debated whether the economic problems of the 1970s were due to the oil supply disruptions or to inappropriate fiscal, monetary, and energy policies implemented to deal with them. The consensus is that the disruptions would have caused economic problems irrespective of fiscal, monetary, and energy policies, but that price and allocation controls exacerbated the impacts in the U.S. during the 1970s.[122] There is general consensus on the following:

- Appropriate actions taken included CAFE, the 55 mph speed limit, reorganization of the Federal energy bureaucracy, greatly increased energy R&D, establishment of the Strategic Petroleum Reserve (SPR), energy efficiency standards and building codes, establishment of IEA and EIA, and burden sharing agreements among nations.
- Inadvisable actions included price and allocation controls, excessive regulations, de-facto gasoline rationing, "excess profits" taxes, policies targeting "greedy energy companies," prohibitions on energy use, and subsidy programs.
- Some actions that seemed to be inappropriate may have been desirable if the problem had not been short-lived. For example, synthetic fuel initiatives may have looked prescient had oil prices not collapsed in the mid 1980s.[123]

Estimated costs to the U.S. of oil supply disruptions range from $25 billion to $75 billion per year, and the cumulative costs since 1973-74 total about $4 trillion.[124] Nevertheless, except for several serious disruptions (and then only temporarily), oil prices have risen little in real terms over the past century, as shown in Figure A-2.

Cost of living adjustment clauses imbedded in many contracts, labor agreements, and government programs (e.g., Social Security) are less visible but important inflation drivers. Price increases generated by oil supply disruptions automatically trigger successive inflationary adjustments throughout the economy, and these complicate monetary policies designed to counter the inflationary effects of the disruption.[125]

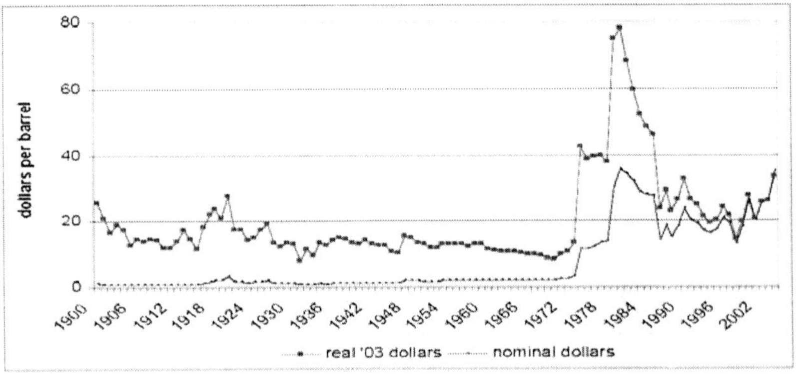

Figure A-2. Oil Prices in Current and Constant Dollars: 1900 - 2004

The U.S. is currently less oil-dependent (in terms of oil / GDP ratios) than during the 1970s. However, as shown in Figure A-3, the U.S. is now importing twice as much oil (in percentage terms) as 30 years ago and its transportation sector consumes a larger portion of total oil consumption.[126] Further, by 2000 most of the energy saving trends resulting from the 1970s disruptions (increased energy efficiency and conservation, increased vehicle mpg, etc.) had been captured.

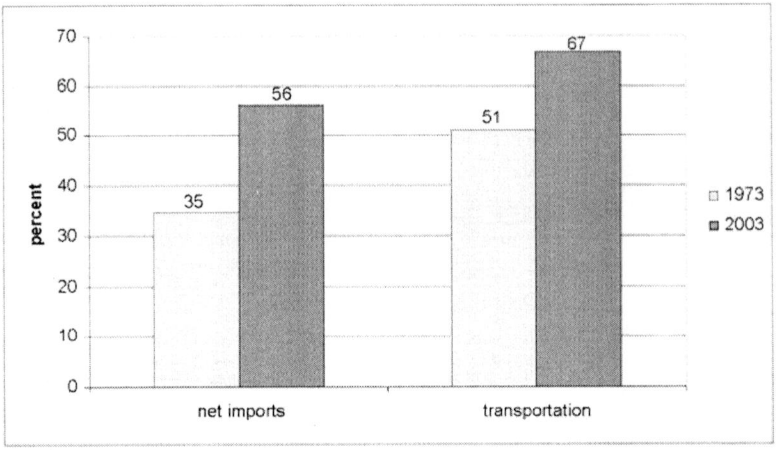

Figure A-3. U.S. Oil Imports and Transportation Shares of Oil Consumption, 1973 and 2003

The primary effect of the 1973-74 disruption was oil price increases. As shown in Figure A-2, the real price of oil peaked in 1981 and has never again reached similar levels.

At present, oil would have to be nearly $80 per barrel and gasoline would have exceed $3 per gallon to equal real 1981 prices. Even then, however, energy would still be less significant factor in the U.S. economy because average U.S. per capita incomes have doubled since 1981 and energy is a much smaller component of expenditures[127].

Nevertheless, over the past 20 years, oil prices have been extremely volatile – more volatile than virtually any other commodity.[128]

APPENDIX III
LIKELY FUTURE OIL DEMAND

Petroleum consumption has been inexorably linked to population growth, industrial development, and economic growth for the past century. This relationship is expected to continue worldwide for the foreseeable future. While the U.S. consumes more oil than any other country – about 20 MM bpd, it represents only 26 percent of world production, compared to the 46 percent of world oil production the U.S. consumed in 1960. As shown in Figure A-4, Western Europe currently consumes the second largest amount (18 percent) followed by Japan (7 percent), China (6 percent), and the FSU (5 percent), with over 150 other countries accounting for the remaining 38 percent of production.[129]

Energy forecasting is difficult due to the numerous complex factors that influence energy supply and demand.[130] Here we utilize the U.S. Energy Department's Energy Information Administration forecasts of future world oil requirements.

Table A-1 presents summary statistics for the EIA 2001-2025 forecast including 24-year country or country group projections for petroleum consumption, gross domestic product (GDP), and population.

Oil consumption in China is expected to increase 4 percent a year, and by 2025 China is projected to be the second largest oil consuming country in the world, accounting for 11 percent of total world consumption. The second fastest growing market is projected to be the FSU countries, where petroleum consumption is forecast to increase an average of over 2 percent per year.

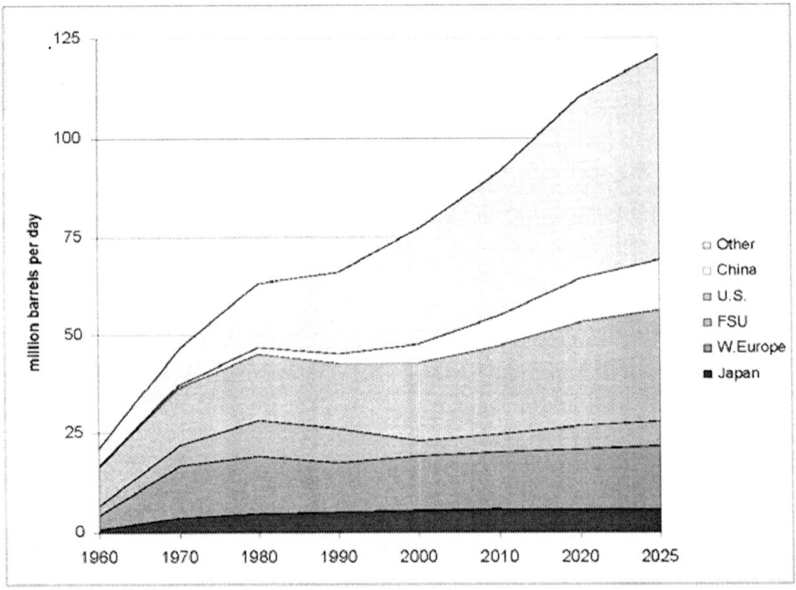

Figure A-4. World Petroleum Consumption, 1960-2025

Table A-1.Reference Case Projections, 2001-2025 (Average annual % change)[131]

	Petroleum Consumption	GDP (Con. $)	Population
U.S.	1.5	3.0	0.8
W.Europe	0.5	2.0	0.1
China	4.0	6.1	0.5
FSU	2.1	4.2	-0.2
Japan	0.3	1.7	-0.1
Other	2.0	4.0	1.3
World	**1.9**	**3.0**	**1.0**

The remaining large consumers, including the U.S., Western Europe, and Japan are forecast to experience consumption growth over the 24-year period at or below the world average. The U.S. is forecast to increase oil consumption at a rate of 1.5 percent per year, and by 2025 the U.S. share of world oil consumption is forecast to decline to 23 percent (29.7 MM bpd), while Western Europe's share decreases to 13 percent (14.4 MM bpd). The many countries grouped as "Other" above, including India, Mexico, and

Brazil, are expected to experience oil consumption growth rates 10 to 30 percent higher than the world average. By 2025, this group is forecast to account for 43 percent of world oil consumption.

In sum, in the EIA reference case, world oil consumption of 80 MM bpd in 2003 is projected to increase to 121 MM bpd in 2025, with the most rapid increases occurring in nations other than the U.S., Japan, or those in Western Europe. Average annual world oil demand growth is projected as 1.9 percent over the period.

APPENDIX IV
RATIONALES FOR THE WEDGES

A. Vehicle Fuel Efficiency

The original U.S. Corporate Average Fuel Efficiency (CAFE) timetable, enacted in 1975, mandated a 53 percent increase in vehicle fuel efficiency, from 18 mpg to 27.5 mpg, over the seven years between 1978 and 1985. Average on-road vehicle fuel efficiency began to improve markedly in the early 1980s and continued to improve substantially every year through 1995. It showed little change between 1995 and 1999, and then began to decline gradually due to the shift to greater purchases of light trucks and SUVs. Between 1982 and 1995, average on-road vehicle fuel efficiency increased from about 14 mpg to 20 mpg. In other words, the first major U.S. oil disruption occurred in the fall of 1973; CAFE was not enacted until two years later; the increased mpg requirements did not begin until 1978, and were phased in through 1985; and significant increases in average on-road vehicle fuel efficiency did not occur until the mid-to late 1980s.[132]

From the time world oil peaking occurs or is recognized, it may thus take as long as 15 years until strengthened vehicle fuel efficiency standards significantly increase average on-road fleet fuel efficiency. However, care must be exercised in making extrapolations. Most "realistic" enhanced vehicle fuel efficiency standards might not actually decrease future total gasoline consumed in the U.S. due to the anticipated continued increase in numbers of drivers and vehicles. Thus, a new CAFE mandate might decrease the rate at which future gasoline consumption increases, but not necessarily reduce total consumption.[133] Only aggressive vehicle fuel efficiency standards legislation that "pushes the envelope" of fuel efficiency technologies over the next two decades (as determined, for example, in the study by the National Research Council of the National Academy of Sciences [134]) is likely to actually reduce total U.S. gasoline consumption.

Savings in the U.S. Assuming a crisis atmosphere, we hypothesize an aggressive vehicle fuel efficiency scenario, based on the NRC CAFE report and other studies that estimate the fuel efficiency gains possible from incremental technologies available or likely to be available within the next decade.[135] We assume that legislation is enacted on the action date in each scenario. We further assume that vehicle fuel efficiency standards are increased 30 percent three years later --for cars from 27.5 mpg to 35.75 mpg and for light trucks from 20.7 mpg to 26.9 --and then increased to 50 percent above the base eight years later --for cars from 27.5 mpg to 41.25 mpg and for light trucks from 20.7 mpg to 31 mpg; finally, we assume full implementation is assumed 12 years after the legislation is enacted. These assumptions "push the envelope" on the fuel efficiency gains possible from current or impending technologies.[136]

On the basis of our assumptions, the U.S. would save 500 thousand barrels per day of liquid fuels 10 ten years after legislation is enacted; 1.5 million barrels per day of liquid fuels at year 15; and 3 million barrels per day of liquid fuels at year 20.

Worldwide Savings. The U.S. currently has about 25 percent of total world vehicle registrations, but consumes nearly 40 percent of the liquid fuels used in transportation worldwide.[137] Since we could not find credible forecasts of the potential impacts of increased worldwide vehicle fuel efficiency standards, we assumed that the impact in the rest of the world of enhanced vehicle fuel efficiency standards will be about equal to that in the U.S. In total, the worldwide impact of increased vehicle fuel efficiency standards would thus yield a savings of 1 million barrels per day of liquid fuels 10 years after legislation is enacted; 3 million barrels per day 15 years after legislation is enacted; and 6 million barrels per day 20 years after legislation is enacted.

Increased vehicle fuel efficiency standards are a powerful way to reduce liquid fuels consumption. However, they required long lead-times to enact, implement, and become effective in the past. On the other hand, their importance and contributions continue to grow over time as older vehicles are retired. Our world vehicle fuel efficiency wedge is assumed to be as follows:

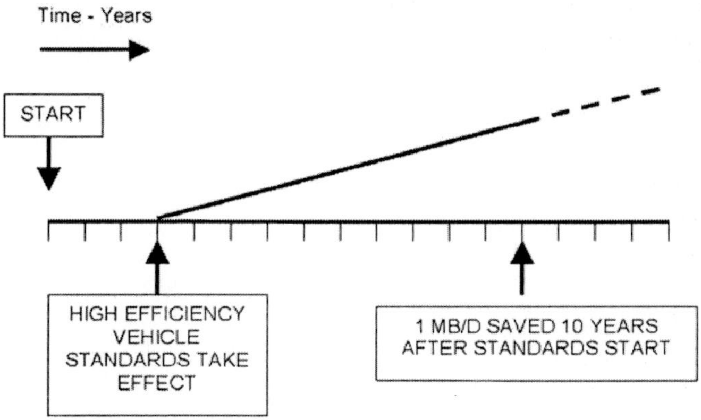

We note that a detailed study of these issues and opportunities would be of great value.

B. Coal Liquids

High quality liquid fuels can be made from coal via direct liquefaction or via gasification followed by Fisher-Tropsch synthesis. A number of coal liquefaction plants were built and operated during World War II, and the Sasol Company in South Africa subsequently built a number of larger, more modern gasification-based facilities.[138]

While the first two Sasol coal liquids production plants were built under normal business conditions, the Sasol Three facility was designed and constructed on a crash basis in response to the Iranian revolution of 1978-79. The project was completed in just over three years after the decision to proceed. Sasol Three was essentially a duplicate of Sasol Two on the same site using a large cadre of experienced personnel. Sasol Three was brought "up to speed almost immediately."[139]

The Sasol Three example represents the lower bound on what might be accomplished in a twenty-first century crash program to build coal liquefaction plants. This is because the South African government made a quick decision to replicate an existing plant on an existing, coal mine-mouth site without the delays associated with site selection, environmental reviews, public comment periods, etc. In addition, engineering and construction personnel were readily available, and there were a number of manufacturers capable of providing the required heavy process vessels, pumps, and other auxiliary equipment. While we have not done a survey of worldwide capabilities to perform similar tasks today, it is our belief that such capabilities are now in much shorter supply – a situation that will worsen dramatically with the advent of a worldwide crash program to build alternate fuels plants. We have therefore attempted to strike a balance between what we believe could be a somewhat slow startup of a worldwide coal liquefaction industry and a later speed up as experience is gained and new plants are built as essentially duplicates of previous plants.

Our coal liquefaction wedge thus assumes that the first coal liquefaction plants in a worldwide crash program would begin operation four years after a decision to proceed. We assume plant sizes of 100,000 bpd of finished, refined product, and we assume that five such plants could be brought into operation each year. We cannot predict where in the world these coal liquefaction plants might be built. Candidate countries with large coal reserves include the U.S. and the Former Soviet Union with the largest, followed in descending order by China, India and Australia.[140] We note that a consortium of Chinese companies has recently signed a letter of intent with Sasol for feasibility studies on the construction of two new coal-to-liquids plans in China.[141]

Our coal liquefaction wedge then appears be as follows:

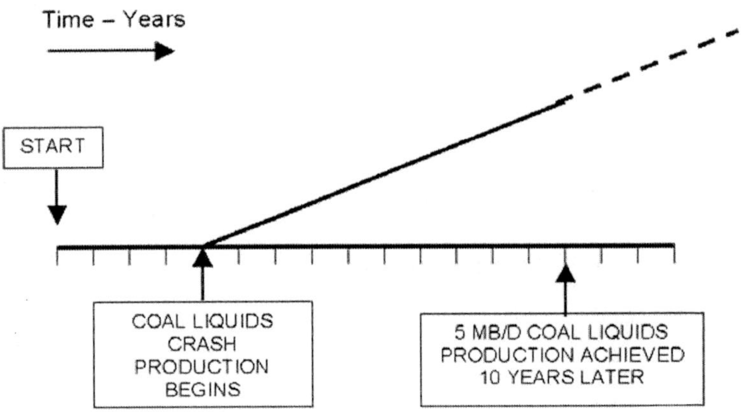

If U.S. siting and environmental reviews of new energy facilities were to continue to be as time consuming as they are today, few coal liquefaction plants would likely be built in the U.S. On the other hand, China has been quick to approve major new facilities, so coal liquefaction plants in that country might well be built expeditiously and economically. Because there is presently a large international trade in coal, it is not inconceivable that coal-poor counties might become the sites of many coal liquefaction plants using imported coal, possibly even from the U.S.

C. Heavy Oils / Oil Sands

As noted, significant heavy oil production currently exists in Canada and Venezuela. While their total resource is estimated to be 3-4 trillion barrels, recoverable oil reserves are estimated to be roughly 600 billion barrels.[142] Such reserves could support a massive expansion in production of these unconventional oils.

In the case of Canadian oil sands, a number of factors would challenge a crash program expansion, such as the need for massive supplies of auxiliary energy, huge land and water requirements, environmental management, and the harsh climate in the region. In the case of Venezuela, large amounts of supplemental energy, inherently low well productivity and other factors will likely pose significant challenges.

We know of no comprehensive analysis of how fast the Canadian and Venezuelan heavy oil production might be accelerated in a world suddenly short of conventional oil. Recent statements by the World Energy Council (WEC) guided our wedge estimates:[143]

- "Unconventional oil is unlikely to fill the gap (associated with conventional oil peaking). Although the resource base is large and technological progress has been able to bring costs down to competitive levels, the dynamics do not suggest a rapid increase in supply but, rather, a long, slow growth over several decades."
- "(Extrapolating expectations of TOTAL Oil Company in the Orinoco, Venezuela) overall reserves today would be only ~60 Gb over 30 years, allowing at best 6 MM bpd of production in 2030 if the entire area were put into production."
- "Current estimates put the additional production of Canada (heavy oil)… at less than 2 MM bpd in 2015-2025."

In line with the WEC, we assume the following for our Venezuelan Heavy Oils wedge:

1. Accelerated production might begin three years after a decision to proceed with a crash program. This delay is based on the fact that

the country already has significant production underway. Starting from scratch would require much more time.

2. Under business-as-usual conditions assumed by the WEC, Venezuela would have production of 6 MM bpd in 2030 --5.5 MM bpd beyond production of 0.5 MM bpd in 2003. If we assume this level of production is achieved 10 years after initiation of a crash program, rather than the roughly 25 years estimated by WEC, then roughly 5.5 MM bpd of incremental production might be achieved 13 years from a decision to accelerate.
3. In contrast to the WEC, we assume that Venezuelan production is not capped at 6 MM bpd but continues to expand for the period covered by our approximations. Note: We ignore the currently extremely unstable political environment in Venezuela and assume that scale-up timing is not hindered by local politics.

Our assumptions for Canadian oil sands are as follows:

1. Again, accelerated production might begin three years after a decision to proceed with a crash program, based in large part on the fact that the country already has significant production underway.
2. Current plans are for production of 3 MM bpd of synthetic crude oil from which refined fuels can be produced by 2030. This is above current production of 0.6 MM bpd. If we assume this level of production is achieved 10 years after initiation of a crash program, rather than the roughly 25 years targeted by the Canadians, then roughly 2.5 MM bpd of incremental production might be achieved 13 years from a decision to accelerate.
3. aWe know of no upper limit on Canadian oil sands production, so for purposes of this order-of-magnitude illustration, we do not assume one.

Our heavy oil wedge therefore is approximated as follows:

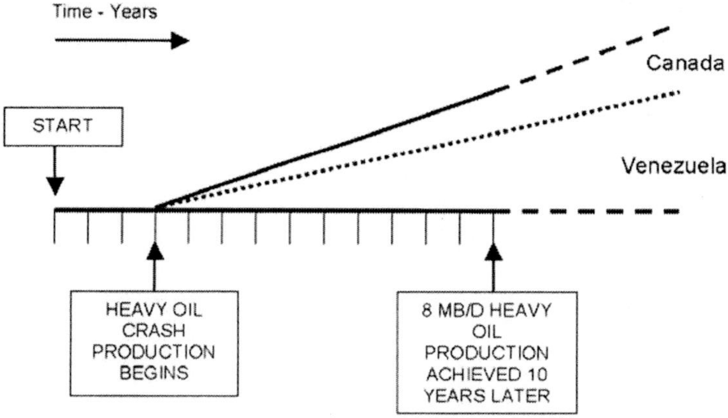

D. Enhanced Oil Recovery

Because it is impossible to evaluate the worldwide impact of Improved Oil Recovery (IOR) techniques, we can only provide a rough estimate of what might be achieved. We focus on a major subset of IOR technologies – Enhanced Oil Recovery (EOR). While EOR can add significantly to reserves, it is normally not applied to a conventional oil reservoir until after production has peaked. As discussed earlier, the most widely applicable EOR process involves the injection of CO2 into conventional oil reservoirs to dissolve and move residual oil. Because EOR processes require extensive planning, large capital expenditures, procurement of very large volumes of CO2, and major equipment for large reservoirs, our simplified assumptions parallel those for our heavy oil and coal liquids wedges.

We assume that the massive application of EOR worldwide will not begin to show production enhancement until 5 years after the peaking of world oil production, paced primarily by the difficulties of procuring CO2. We further assume that world oil production enhancement due to such a crash effort worldwide will increase world oil production by roughly 3 percent after 10 years.[144] We translate the 3 percent to 3 MM bpd, based on our assumed world oil peaking level of roughly 100 MM bpd. Our EOR wedge thus appears as follows:

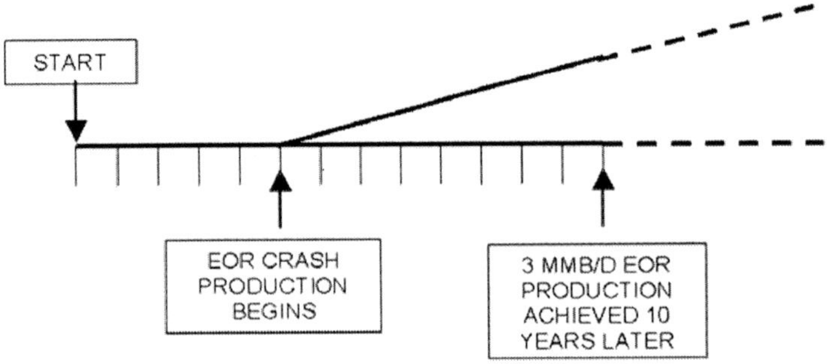

E. Gas-To-Liquids

Estimating how fast world Gas-To-Liquids (GTL) production might grow as a result of the peaking of world oil production is an extremely complex undertaking because of the need to consider the total world energy system, its likely growth by country, future energy economics, other resources that compete with natural gas, etc. In a crash program, GTL plants might be built in a number of counties that have large reserves of stranded gas.. Once operational, GTL product could be moved to markets around the world by conventional oil product tankers.

Our estimates for a crash program of world GTL production are tempered by the conflicting world demand for Liquefied Natural Gas (LNG), whose export volumes are currently growing at a rapid pace. The tradeoffs involved in estimating the future LNG / GTL balance are complex, and a world crash program in GTL could yield higher or lower volumes than our estimates. Note also that seven countries currently account for almost 80 percent of the world gas export market, and it is not inconceivable that the recently formed Gas Exporting Countries Forum (GECF) might well evolve into a future OPEC-like cartel.[145]

Again, we assume a startup delay of three years before crash program GTL plants might come into operation. Using a base case, business-as-usual production forecast of 1.0 MM bpd in 2015 from the current level of essentially zero, we assume that a crash program might yield the 1.0 MM bpd in 5 years. The resultant wedge might then be as follows:

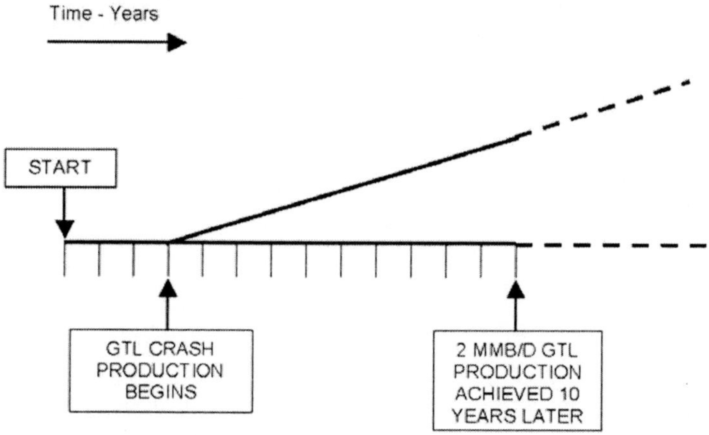

F. Sum of the Wedges

A summary of the estimates from the foregoing is presented in Table A-2.

Table A-2. Summary of Consumption and Production Wedge Estimates

Category	Delay until first Impact (Years)	Impact 10 Years Later (MM bpd)
Vehicle Efficiency	3	3
Gas-To-Liquids	3	2
Heavy Oils / Oil Sands	3	8
Coal Liquids	4	5
Enhanced Oil Recovery	5	3

Ordering the various contributions by their starting dates, the total mitigation wedge is as shown in Figure A-5.

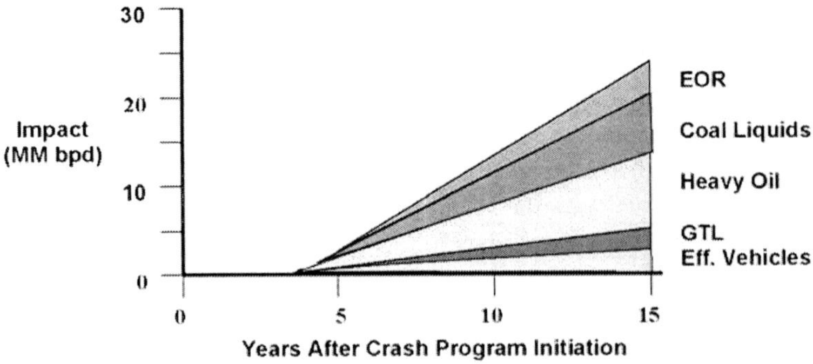

Figure A-5. The total of the wedge estimates

Appendix V
Notes on Shale Oil and Biomass

A. Oil Shale by Gilbert McGurl, NETL

Worldwide resources of oil shale comprise an estimated 2.6 trillion barrels, of which two trillion are located within the United States. The richest deposits, 1.5 trillion bbl with high concentrations of kerogen, lie in Colorado, Utah, and Wyoming. An additional 16 billion barrels of rich but physically different oil shale is found in Kentucky, Indiana, and Ohio. A recent estimate is that, from the Green River deposits, 130 billion barrels of oil may be produced. Technology development on oil shale 'retorting' reached a high point in the late 1970s, with the major oil companies leading the way. The oil price collapse of the 1980s, the dissolution of the synfuels program, and the termination of the Unocal project in 1991 led to the demise of oil shale production in the United States.

A recent study performed by the DOE Office of Naval Petroleum and Oil Shale Reserves advocates a research and development program with a production goal of two million barrels per day by 2020.[146] Production would be initiated by 2011. Traditional technologies for mining and preparation of oil shale ores and for aboveground upgrading have been 'proven' at less-than-commercial scale. Newer Canadian technologies have been tested at demonstration projects in Australia. However, that project, the Stuart upgrading project, is currently suspended pending project re-design. Nonetheless, the same technology has been licensed by operators in Estonia.

Technologies for in-situ recovery are newer and less developed. In 2000, Shell revived an oil shale project called "Mahogany" in Colorado.[147] Shell aims to test its process until 2010. If successful, the in-situ method would leave heavier hydrocarbons in the shale while producing lighter hydrocarbons and using much less water than traditional methods.

Most Estonian processing of oil shale has been for boiler fuel for electricity production. Small liquids facilities have been operating at "full capacity" given recent market oil prices. There are no solid figures for cost in large-scale plants since none have been built. The aborted Australian project estimated $8.50/bbl in operating costs once a commercial plant had been built. The Estonians estimate a break-even point at $21 Brent price (app $23 WTI) and low capacity factor. At higher capacity factors, plants may operate profitably even with prices in the mid-teens.

Besides water use and production, environmental concerns include fine particulates and carbon dioxide emissions. Since the last US oil shale project ceased operation before the implementation of the 1990 Clean Air Act amendments, new emission-control equipment would need to be tested on US shales.

B. Biofuels by Peter Balash, NETL

Bioethanol is produced as a transportation fuel largely in only two countries. In 2003 the US produced about 2.8 billion gallons and Brazil produced 3.5 billion gallons. All of this ethanol is produced by conversion of starch to sugar and fermentation to ethanol. In the US ethanol represents about 1.4% of the BTU content (2.0% by volume) of gasoline used in transportation. Current costs for ethanol production in the US are said to be $0.90 per gallon, [148] which is equivalent to a gasoline price of $1.35 per gallon. Because of recent increases in energy costs current costs will be somewhat higher. Grain ethanol provides only a modest net energy gain because of the energy required to produce it. USDA calculated a net energy gain of 34% for a modern corn to ethanol plant, [149] but there is considerable controversy over the real efficiency of the process. Most of the energy used to produce ethanol comes from natural gas and electricity. The production of ethanol uses only about 5% of the corn crop in the US. Significant expansion is possible but at some point there might be an impact on food prices.

Cellulosic ethanol is currently being produced only in two rather small pilot plants but is capable of producing about 40% conversion of cellulosic biomass to ethanol while providing all the energy needed for the process and exporting a modest amount of energy as electricity. It is anticipated that successful research may reduce the cost of cellulosic ethanol to about $1.10 per gallon by 2010. If this occurs the potential ethanol to mitigate peaking is high. Using only waste biomass and grass grown on land currently in the conservation reserve could produce 50 billion gallons of ethanol which would be equivalent to 35 billion gallons of gasoline or 17% of current US consumption. This could be achieved without any impact on current food production and at prices only $ 0.35 per gallon higher than refinery prices for gasoline. Since ethanol has an RON of 130 and a MON of 96 it raises the octane of the gasoline to which it is added and has a premium value as a result.

APPENDIX VI
AREAS FOR FURTHER STUDY

1. Economic Benefits to the U.S. Associated with an Aggressive Mitigation Initiative

Important economic and jobs benefits could result from a concerted U.S. effort to develop substitute fuels plants based on U.S. coal and shale resources and scale up of EOR. The impacts might include hundreds of billions of dollars of investment, hundreds of thousands of jobs, a rejuvenation of various domestic industries, and increased tax revenues for the Federal, state, and local governments. The identification and analysis of such benefits require analysis.

In the short run, the U.S would be hard-pressed to find adequate physical and human resources to plan, develop, construct, and operate the required facilities. Given that oil peaking is a world problem, it is virtually certain that at the same time the U.S. embarked on an aggressive mitigation program, other major initiatives would likely be undertaken elsewhere in the world. All would require similar types of capital, technology, and human resources, generating additional constraints and inflationary pressures on the U.S. program. Assessment of the impacts of these constraints on the feasibility, costs, and timing of a major U.S. mitigation program merits investigation.

2. Oil Peaking Risk Analysis: Cost of Premature Mitigation versus Waiting

The date of world oil production peaking is unknowable, but it may occur in the not too distant future. Large-scale mitigation is needed more than a decade before the onset of peaking if economic hardship is to be avoided. If major efforts were initiated early and peaking was to occur decades later, there might be an unproductive use of resources. On the other hand, mitigation initiated at the time of peaking will not spare the world from a decade or more of devastating economic impacts. A careful analysis of the benefits / costs of early versus late mitigation could provide valuable insights.

3. U.S. Natural Gas Production as a Paradigm for Viewing World Oil Peaking

The history of U.S. natural gas production is cited as an example of the perils of over-optimistic resource forecasts. A detailed analysis of the North American natural gas history, status, and outlook might provide lessons useful in addressing world oil production peaking.

4. Potential for Non-transportation Oil Fuel-Switching

World non-transportation liquid fuel usage is amenable to fuel switching, thereby freeing up liquids for transportation. If switching were to occur on a large-scale, it would likely take place gradually because other energy substitutes would have to be scaled up to meet the new demands associated with a major shift, e.g., electric power plants built, refineries expanded to produce a different product slate, etc. A detailed study would provide an understanding of how difficult, expensive, time-consuming and productive worldwide non-transportation fuel switching might be.

5. World Coal-To-Liquids Potential

Sasol has operational coal-to-liquids (CTL) production plants and is under contract to study the construction of similar facilities in China. An analysis of worldwide large-scale CTL potential could yield a useful estimate of complexity, timing and potential.

6. World Heavy Oil / Oil Sands Potential

Canada, Venezuela, and, to a lesser degree, other countries have potential to massively scale up their unconventional oil production. A better understanding of how quickly scale-up might be implemented, the related barriers, and ultimate potential would help in the understanding the potential contribution of these resources.

7. World EOR Potential

An analysis of worldwide large-scale EOR potential could provide an estimate of complexity, timing and potential.

8. World GTL Potential

An analysis of worldwide large-scale GTL potential could yield a useful estimate of complexity, timing and potential. In particular, the likely conflicts between GTL and LNG production could provide a quantitative estimate of likely future use of world stranded gas.

9. World Transportation Fuel Efficiency Improvement Potential

It is important that we have the best possible understanding of the U.S. and worldwide potential for the upgrading of transportation fuel efficiency, including possible timing, cost, and savings as a function of time. Excellent data is available on U.S. transportation fleets, but fleets elsewhere in the world are less well described. A careful study is needed.

10. Impacts of Oil Prices and Technology on U.S. Lower 48 Oil Production

Analysis of U.S. Lower 48 oil production since the 1970 peak strongly suggests that oil prices and advancing technology had little impact on the production decline. However, a number of institutional factors also impacted Lower 48 oil production, e.g., allowables (Texas Railroad Commission), price and allocation controls (1970s), free market pricing (since 1981), foreign opportunities for multi-national oil companies, etc. An in-depth understanding of these various influences might provide useful guidance for the future.

11. Technological Options for Coal Liquefaction

Current world coal liquefaction R & D is focused on gasification of coal followed by the Fischer-Tropsch synthesis. Other coal-to-liquids processes have been proposed, some of which were tested at relatively large scale. It may be worthwhile to revisit the various options in light of today's technology and environmental requirements to determine if any of them might also have competitive potential.

12. Performance of Oil Provinces Outside of the U.S.

There is a strong rationale for using U.S Lower 48 oil production as a surrogate pattern for future world oil production peaking and decline. Other large oil province histories could also yield valuable insights and alternate patterns. Related analysis might provide an improved basis for modeling future world oil production.

13. How the U.S. Could Again Become the World's Largest Oil Producer

After the peaking of world conventional oil production, there will be a major world transition from the current world liquid fuel infrastructure. Over time, major conservation and energy switching initiatives will almost certainly be implemented, but the need for liquid fuels will not disappear for at least the remainder of this century because there are no known alternatives for a number of transportation applications. An analysis of the major factors required for the U.S. to return to a position of oil supremacy and oil independence would be enlightening.

14. Market Signals in Advance of Peaking

Increases in oil prices and oil price volatility have been identified as two precursors of world oil peaking, but both are likely short-term signals. The identification and character of longer-term signals, if they exist, could be of significant value.

15. Risk of Repeating the Synthetic Fuels Experience of 1970s and 1980s

One risk of embarking on aggressive oil peaking mitigation is that OPEC might undermine such efforts by dramatically increasing conventional oil production. This could only happen if excess capacity were to exist, which could happen if world oil peaking was many decades away. Were such a dramatic increase in OPEC production to occur, governments would be under pressure to terminate support for their mitigation programs. Related scenarios might worthy of study.

16. Effects of Oil Price Spikes in Causing U.S. Recessions

Oil price spike have been followed by U.S. recessions, but they are not the only cause of recessions. A detailed study of the role of oil prices and other factors in causing recessions might be worth further study.

REFERENCES

[1] In this study we interchangeably refer to the peaking of world conventional oil production as "oilpeaking" or simply as "peaking."

[2] Portions of this chapter are taken from Hirsch, R.L. "Six Major Factors in Energy Planning". U.S. Department of Energy. National Energy Technology Laboratory. March 2004.

[3] U.S. Department of Energy, Energy Information Administration, *International Energy Outlook* – 2004, April 2004.

[4] U.S. Department of Energy, Energy Information Administration, *International Energy Outlook* –2004, April 2004.[6]

[5] Total oil in place is called the "resource." However, only a part of the resou rce can be produced, because of geological complexities and economic limitations. That which is realistically recoverable is called "reserves," which varies within limits depending on oil prices.

[6] Aleklett, K. & Campbell, C.J. "The Peak and Decline of World Oil and Gas Production". Uppsala University, Sweden. ASPO web site. 2003.

[7] U.S. Department of Energy, Energy Information Administration, *International Energy Outlook* – 2004, April 2004.

[8] U.S. Department of Energy, Energy Information Administration, Long Term World Oil Supply, April 18, 2000.

[9] The US Lower 48 experience occurred over a long period characterized at different times by production controls (Texas Railroad Commission), price and allocation controls (1970s), free market prices (since 1981), wild price swings, etc., as well as higher prices and advancing technology. Nevertheless, production peaked and moved into a relatively constant rate of decline.

[10] Bakhtiari, A.M.S. "World Oil Production Capacity Model Suggests Output Peak by 2006-07."OGJ. April 26, 2004.

[11] Simmons, M.R. ASPO Workshop. May 26, 2003.

[12] Skrebowski, C. "Oil Field Mega Projects -2004." Petroleum Review. January 2004.

[13] Deffeyes, K.S. Hubbert's Peak-The Impending World Oil Shortage. *Princeton University Press*.2003.

[14] Goodstein, D. Out of Gas – The End of the Age of Oil. W.W. Norton. 2004

[15] Campbell, C.J. "Industry Urged to Watch for Regular Oil Production Peaks, Depletion Signals."OGJ. July 14, 2003.

[16] Drivers of the Energy Scene. World Energy Council. 2003.

[17] Laherrere, J. Seminar Center of Energy Conversion. Zurich. May 7, 2003

[18] DOE EIA. "Long Term World Oil Supply." April 18, 2000. See Appendix I for discussion.

[19] Jackson, P. et al. "Triple Witching Hour for Oil Arrives Early in 2004 – But, As Yet, No RealWitches." *CERA Alert*. April 7, 2004.

[20] Davis, G. "Meeting Future Energy Needs." The Bridge. *National Academies Press*. Summer2003.

[21] Lynch, M.C. "Petroleum Resources Pessimism Debunked in Hubbert Model and HubbertModelers' Assessment." *Oil and Gas Journal*, July 14, 2003.

[22] U.S. Department of Commerce, Bureau of Economic Analysis, National Income and ProductAccounts, 2004.

[23] U.S. Department of Transportation, Federal Highway Administration, Highway Statistics, 2004.

[24] U.S. Department of Energy, Energy Information Administration, *Monthly Energy Review*, 2004.

[25] U.S. Department of Energy, Energy Information Administration, Detailed annual petroleum consumption accounts by fuel and sector at www,eia.doe,gov, 2004

[26] Because of the lack of national average "replacement value" estimates, current-cost net capital stock provides a suitable substitute for the estimates. Given the capital equipment depreciation schedule used, the total replacement value of the capital stock is projected to be 4.5 times higher than the current-cost net value

[27] U.S. Department of Energy, Energy Information Administration, Annual Energy Outlook -2004, and Oak Ridge National Laboratory, *Transportation Energy Data Book* #23, 2003.

[28] U.S. Department of Commerce, Bureau of Economic Analysis, Fixed Asset Tables, 1992-2002. The estimate of net stock includes an adjustment for depreciation, defined as the decline in value of the stock of assets due to wear and tear, obsolescence, accidental damage, and aging. For most types of assets, estimates of depreciation are based on a geometric decline in value.

[29] Oak Ridge National Laboratory, *Transportation Energy Data Book* #23, 2003; and U.S. Department of Transportation, Bureau of Transportation Statistics, Active Air Carrier Fleet; and Management Information Services, Inc., 2004.

[30] The largest remaining oil-consuming capital stock resides in the industrial sector. Oil consumption in the industrial sector is diverse, making it difficult to target specific capital stock and identify potential efficiency efforts or potential technology advancements. The largest oil-consuming industries include the chemical, lumber and wood, paper products, and petroleum industry itself. Functional usage of oil in the industry includes heat, process heat, power, feedstock, and lubrication. Finally, the equipment spans hundreds of disparate types of in situ engines, turbines, and agricultural, construction, and mining machinery.

[31] U.S. Department of Transportation, Bureau of Transportation Statistics, American Travel Survey Profile and Oak Ridge National Laboratory, *Transportation Energy Data Book* -2003.

[32] U.S. Department of Energy, Energy Information Administration. *International Energy Annual*, 2004. April 2004.

[33] U.S. Department of Energy, Energy Information Administration, "Latest Oil Supply Disruption Information," eia.doe.gov, 2004; U.S. Department of Energy, Energy Information Administration,. "World Oil Market and Oil Price Chronologies: 1970-2003," March 2004; U.S. Department of Energy, Energy Information Administration, "Global Oil Supply Disruptions Since 1951", 2001; U.S. Department of Energy, Energy Information Administration, Annual Energy Review, 2002;U.S. Department of Energy, Energy Information Administration, *International Petroleum Monthly*, April 2004.

[34] This is verified by the extensive literature review conducted by Donald W. Jones and Paul N. Leiby, "The Macroeconomic Impacts of Oil Price Shocks: A Review of the Literature and Issues," Oak Ridge National Laboratory, January 1996, and by Donald W. Jones, Paul N. Leiby, and Inja K Paik, "Oil Price Shocks and the Macroeconomy: What Has Been Learned Since 1996, *The Energy Journal*, 2003.

[35] See, for example, Leonardo Maugeri, "Oil: Never Cry Wolf – Why the Petroleum Age is Far From Over, " Science, Vol. 304, May 21, 2004, pp. 1114-1115; Michael C. Lynch, "Closed Coffin: Ending the Debate on 'The End of Cheap Oil,' A Commentary," DRI/WEFA, September 2001; Michael C. Lynch "Farce This Time: Renewed Pessimism About Oil Supply, 2000; Bjorn Lomborg, "Running on Empty?" Guardian, August 16, 2001; Mark Mills, "Stop Worrying About Oil Prices," 2001, fossilfuels.org; Jerry Taylor, "Markets Work Magic," Cato Institute, January 2002; Rethinking Emergency Energy Policy, U.S. Congressional Budget Office, December 1994.

[36] This is the consensus of virtually every rigorous analysis of the problem; see, for example, the International Monetary Fund study conducted by Benjamin Hunt, Peter Isard, and Douglas Saxton, "The Macroeconomic Effects of Oil Price Shocks," National Institute *Economic Review* No. 179, January 2002.

[37] "The Impact of Higher Oil Prices on the World Economy," OECD Standing Group on Long-TermCooperation, 2003.

[38] See Lee, Ni, and Ratti, op. cit., and J.D. Hamilton and A.M. Herrera "Oil Shocks and AggregateMacroeconomic Behavior: The Role of Monetary Policy," *Journal of Money*, Credit and Banking,2003.

[39] U.S. Joint Economic Committee and Management Information Services, Inc., 2004.

[40] This totals about $1.1 trillion in 2003 dollars and was equivalent to a once-and-for-all reduction in real GDP of about seven percent; however, part of that loss was likely attributable to structural and cyclical economic factors unrelated to the oil-price shock. See Faith Bird, "Analysis of the Impact of High Oil Price on the Global Economy," International Energy Agency, 2003.

[41] These losses totaled about $700 billion and $1.1 trillion, respectively in 2003 dollars. Losses of this magnitude are significant and represent the difference between vibrant, growing economies and economies in deep recession. There is considerable debate as to precisely how much of these losses was attributable to the oil price shocks, to fiscal and monetary policies, and to other factors.

[42] See Bird, op. cit., and OECD Standing Group on Long-Term Cooperation, op. cit.

[43] A $10/bbl. increase in oil prices, if sustained for a year, will reduce global GDP by 0.6 percent,ignoring the secondary effects on confidence, stock markets, and policy responses; see Bird, op.cit. A sustained increase of $10/bbl. would reduce economic growth by 0.5

percent in the industrialized countries and by 0.75 percent or more in the developing countries; see Ibid., OECD Standing Group on Long-Term Cooperation, op. cit., and International Monetary Fund, World Economic Outlook, September 2003. Larger oil price increases will have even more severe economic effects.

[44] K.A. Mork, "Business Cycles and the Oil Market," *Energy Journal*, special issue, 1994, pp. 15-38.

[45] See Mark Hooker, "Are Oil Shocks Inflationary? Asymmetric and Nonlinear SpecificationVersus Changes In Regime*," Federal Reserve Board*, December 1999.

[46] Hillard Huntington, "Energy Disruptions, Interfirm Price Effects, and the Aggregate Economy,"Energy Modeling Forum, Stanford University, September 2002; S.J. Davis, and J. Haltiwanger, "Sectoral Job Creation and Destruction Response to Oil Price Changes," *Journal of Monetary Economics,* Vol. 48, 2001, pp. 465-512.

[47] "Demand destruction" has often been identified as a solution, since oil price increases resulting from a disruption will reduce demand and this will moderate further price increases. However, demand is reduced because the economy is devastated and large numbers of jobs are lost. Demand destruction – a polite word for economic and job losses – is the problem, not the solution. See the discussion in Roger Bezdek and Robert Wendling, "The Case Against Gas Dependence," *Public Utilities Fortnightly*, Vol. 142, No. 4, April 2004, pp. 43-47.

[48] Joint Economic Committee of the U.S. Congress, "10 Facts About Oil Prices," March 2003; Mark Hooker, "Oil and the Macroeconomy Revisited," *Federal Reserve Board*, August 1999.

[49] Nevertheless, during disruptions, public actions may be required to address societal risks. This creates a dilemma: In the event of a severe shortfall of long duration, government intervention of some sort may be required, and allocation plans to moderate the effects of this shortfall will likely be advocated. However, given the experience of the 1970s, many of the policies enacted in a crisis atmosphere will be, at best, sub-optimal. For example, in 1980, the Federal government developed a Congressionally-mandated stand-by U.S. gasoline rationing plan which could, in some form, be implemented; see Standby Gasoline Rationing Plan, U.S. Department of Energy, Washington, D.C., June 1980.

[50] National Petroleum Council. Meeting the Challenges of the Nation's Growing Natural GasDemand. December 1999.

[51] Esser, R. et al. Natural Gas Productive Capacity Outlook in North America -How Fast Can It Grow? *Cambridge Energy Research Associates*, Inc. 2001.

[52] U.S. Department of Energy, Energy Information Administration, *Annual Energy Outlook* 2000.December 1999.

[53] CERA Advisory Services. The Worst is Yet to Come: Diverging Fundamentals Challenge the North American Gas Market. *Cambridge Energy Research Associates*, Inc. Spring 2004.

[54] Industry Trends (quoting Raymond James & Associates). OGJ. June 7, 2004.

[55] Adkins, J.M. et al. "Energy Industry Brief". Raymond James & Associates. May 17, 2004.

[56] "Lehman Says US 1Q Gas Production Fell By 5.3%". Dow Jones. May 12, 2004.

[57] National Petroleum Council. Balancing Natural Gas Policy – Fueling the Demands of a GrowingEconomy: Volume I – Summary of Findings and Recommendations. September 25, 2003.

[58] Reuters. "Alberta Gas Output Falling Despite Record Drilling". June 6, 2004.

[59] Natural Gas Markets and EIA's Information Program March 2000.

[60] U.S. Department of Energy, Energy Information Administration, Natural Gas Annual 2002.

[61] U.S. Department of Energy, Energy Information Administration, "Natural Gas Navigator." LastUpdated 5/6/04.

[62] CERA Advisory Services. "The Worst is Yet to Come: Diverging Fundamentals Challenge theNorth American Gas Market". Cambridge *Energy Research Associates*, Inc. Spring 2004.

[63] The Alaska natural gas pipeline is at least 10 years from operation, maybe longer.

[64] Flalka, J.J. & Gold, R. "Fears of Terrorism Crush Plans For Liquefied-Gas Terminals." *TheWall Street Journal*. May 14, 2004.

[65] Bender, B. "DistriGas Contests Hazard Study Findings." Boston Globe. June 2, 2004.

[66] Harvan, R. "Diesel Use Surging". *World Refining*. June 2004.

[67] Consumer Reports. August 2004. Page 49.

[68] National Research Council. The Hydrogen Economy: Opportunities, Costs, Barriers, and R & DNeeds. *National Academy Press*. 2004.

[69] Kerwin, K. "Chrysler Puts Some Muscle on the Street". *Business Week*. June 7, 2004.

[70] Press release. Eaton Corp., March 30, 2004.

[71] Press release. National Renewable Energy Technology Laboratory, February 8, 2002.

[72] Homes, S. "A Silver Lining for Boeing". *Business Week.* May 24, 2004.

[73] Williams, B. "Progress in IOR technology, economics deemed critical to staving off world's oil production peak". OGJ. August 4, 2003.

[74] Williams, B. "Progress in IOR technology, economics deemed critical to staving off world's oil production peak". OGJ. August 4, 2003; National Research Council. Fuels to Drive Our Future. *National Academy Press.* 1990.; "EOR Continues to Unlock Oil Resources". OGJ. April 12, 2004.

[75] Economists will argue that this amount will increase with higher world oil prices, which is almost certainly correct. However, without careful analysis, estimation of the increased reserves would be strictly speculation.

[76] These numbers are subject to revision upwards or downwards depending on future geological findings, advancing technology, or higher oil prices. Williams, B. "Heavy Hydrocarbons Playing Key Role in Peak Oil Debate, Future Supply". OGJ. July 28, 2003.

[77] Gray, D. "Oil Sands Conference Report". Mitretek. May 24, 2004.

[78] "Oil Sands Technology Roadmap". Alberta Chamber of Resources. January 2004.

[79] Gray, D. "Oil Sands Conference Report". Mitretek. May 24, 2004.

[80] Underground steam recovery requires about 3 bbls of water per barrel of recovered bitumen.Mining operations need 4-6 bbls of water per bbl of bitumen. Ref.: Gray, D. Oil SandsConference Report. Mitretek. May 24, 2004.

[81] "Oil Sands Technology Roadmap". Alberta Chamber of Resources. January 2004.

[82] Stott, J. "CERI: Alberta Oil Sands Industry Outlook 'Very Robust.'" OGJ. March 22, 2004.

[83] Jaremko, G. "Green forces rally to divert oil sands' use of Arctic gas. Gas use by 2015 couldsurpass Mackenzie capacity". *The Edmonton Journal. April* 15, 2004.

[84] U.S. Department of Energy, Energy Information Administration, "Country Analysis Briefs –Venezuela," June 2004.

[85] Gray, D. "Oil Sands Conference Report". Mitretek. May 24, 2004.

[86] Sen, C.T. "World's LNG Industry Surges, Pushed By Confluence of Factors". June 14, 2004.

[87] Higgins,T. "Gas-To-Liquids: An Emerging Driver for Diesel Markets?" World Refining. April2004.

[88] Kruger, P du P. "Startup Experience at Sasol's Two and Three". Sasol. 1983.

[89] National Research Council. Fuels to Drive Our Future. *National Academies Press*. 1990.

[90] Gray, D. et al. "Coproduction of Ultra Clean Transportation Fuels, Hydrogen, and Electric Power from Coal". Mitretek Systems Technical Report MTR 2001-43, July 2001.

[91] Johnson, H. et al. "Strategic Significance of America's Oil Shale Resource". DOE. March2004.

[92] O'Conner, T. "Mahogany Research Project: Technology to Secure Our Future". *Presentation at the DOE Shale Peer Review*. February 19-20, 2004.

[93] Smith, S.J. et al. "Near-Term US Biomass Potential." PNWD-3285. Battelle Memorial Institute. January 2004.

[94] American Association of Railroads. Railroad Facts. 2002.

[95] "DOE Hydrogen Posture Plan". www.eere.energy.gov/ hydrogenandfuelcells. March 10, 2004.

[96] National Research Council. The Hydrogen Economy: Opportunities, Costs, Barriers and R & DNeeds. National Academies Press. 2004.

[97] Ibid.

[98] There has been extensive discussion of these problems in the literature; see, for example, Management Information Services, Inc., Summary of the Implications of the Environmental Justice Movement for EPRI and its Members; prepared for the Electric Power Research Institute, 1997; K.A Kilmer, G. Anandalingam, and J. Huber, "The Efficiency of Political Mechanisms for Siting Nuisance Facilities: Are Opponents More Likely to Participate Than Supporters?" *Journal of Real Estate Finance and Economics*, vol. 22, 2001; Sheila Foster, "Justice from the Ground Up: Distributive Inequalities, Grassroots Resistance, and the Transformative Politics of the Environmental Justice Movement," California Law Review, vol. 86, no. 4 (1998), pp. 775-841; D. Minehard and Z. Neeman, "Effective Siting of Waste Treatment Facilities," *Journal of Environmental Economics and Management*, vol. 43, 2002, pp. 303-324; Joanne Linnerooth-Bayer, "Fair Strategies for Siting Hazardous Waste Facilities," International Institute for Applied Systems Analysis, Laxenburg, Austria, May 1999; Don Markley, "Its not NIMBY Anymore, its BANANA," Broadcast Engineering, March 1, 2002; S. Tierney and P. Hibbard,

"Siting Power Plants in the New Electric Industry Structure: Lessons From California," The Electric Journal, 2000, pp. 35-49; Dan Sandoval, "The NIMBY Challenge," Recycling Today, April 14, 2003; Philip Sittleburg, "NIMBY Mindset Looks for Zoning Loopholes," Fire Chief, February 1, 2002.

[99] Siting the Brightwater Treatment Facilities: Site Selection and Screening Activities, King County, March 2001.

[100] U.S. Department of Energy, Environmental Siting Guide, Office of Energy Efficiency and Renewable Energy, 2004.

[101] On the other hand, even in the midst of the energy crisis, the Alaska oil pipeline was approved by only one vote in the U.S. Senate and, currently, EIA anticipates that an Alaska gas pipeline will not be completed prior to 2020 – see U.S. Energy Information Administration, 2004 *Annual Energy Outlook*, February, 2004.

[102] U.S. Department of Energy, Energy Information Administration, *Monthly Energy Review*, 2004.

[103] U.S. Department of Energy, Energy Information Administration. "Table 1.2 World Petroleum Consumption, 1980-2002" database and "Table G.2 World Production of Crude Oil, NGPL, Other Liquids, and Refinery Processing Gain 1980-2002" database, 2004.

[104] Ibid

[105] Pacala, S., Socolow, R. "Stabilization Wedges: Solving the Climate Problem for the Next 50 Years with Current Technologies." *Science*. August 13, 2004.

[106] These potential savings are documented in National Research Council, National Academy of Sciences, Effectiveness and Impact of Corporate Average Fuel Economy (CAFE) Standards, Washington, D.C.: *National Academy Press*, 2002; Management Information Services, Inc., and 20/20 Vision, Fuel Standards and Jobs: Economic, Employment, Energy, and Environmental Impacts of Increased CAFE Standards Through 2020, report prepared for the Energy Foundation, San Francisco, California, July 2002; David L. Greene and John DeCicco, Engineering-Economic Analysis of Automotive Fuel Economy Potential in the United States, paper presented at the IEA International Workshop on Technologies to Reduce Greenhouse Gas Emissions, Washington, D.C., May 1999; David Friedman, et al, Drilling in Detroit: Tapping Automaker Ingenuity to Build Safe and Efficient Automobiles, Union of Concerned Scientists, UCS Publications, Cambridge, MA, June 2001; Roland Hwang, Bryanna Millis, and Theo Spencer, Clean Getaway: Toward Safe and Efficient

Vehicles, Natural Resources Defense Council: New York, July 2001; Brent D. Yacobucci, Marc Ross, Technical Options for Improving the Fuel Economy of U.S. Cars and Light Trucks by 2010-2015, American Council for an Energy Efficient Economy, July 2001; Robert L Bamberger, Automobile and Light Truck Fuel Economy: Is CAFE Up to Standards? Washington, D.C.: Congressional Research Service, September 29, 2001; Energy and Environmental Analysis, Inc. Technology and Cost of Future Fuel Economy Improvements for Light-Duty Vehicles, prepared for the National Research Council, 2001.

[107] The choice of a minimum is subjective. A minimum of 50 percent seems reasonable, but ahigher rate is clearly more desirable.

[108] While diesel engines offer significant improvements in fuel economy over gasoline engines,their benefits are notably less than hybrids. For simplicity, we neglect the broader use of dieselsin this study, which is not meant to imply that they might indeed make an important contribution inthe LDV markets.

[109] In their recently published hydrogen study, the National Research Council has shown that hydrogen from biomass is roughly three times as expensive as coal-based hydrogen. This relationship holds roughly for liquids production, another basis for not considering biomass fuels as acceptable under our criteria. See National Research Council, National Academy of Sciences, The Hydrogen Economy: Opportunities, Costs, Barriers, and R&D Needs, Washington, D.C.: *National Academy Press*, 2004

[110] "Hydrocarbon Resources: Future Supply and Demand." World Energy Council -18 th Congress, Buenos Aires, October 2001.

[111] Compounding starts at 67.3 MM bpd at –20 years, rises to 100 MM bpd at year 0, and dropsto 66.8 MM bpd at +20 years.

[112] See for instance Al-Husseini, S.I. , Retired Exec. V.P., Saudi Aramco. A Producer'sPerspective on the Oil Industry. Oil and Money Conference. London. October 26, 2004;Hakes, J. Long Term World Oil Supply. EIA. April 18, 2000; and ExxonMobil. A Report onEnergy Trends, Greenhouse Emissions and Alternate Energy. February 2004.

[113] Over the past 20 years, oil prices have been extremely volatile. Between 1982 and 2002, thestandard deviation in monthly oil prices was 29.5 percent of its mean. The only other majorcommodity whose price exhibited similar volatility was coffee – 27.8 percent of its mean. SeeAndre Plourde and G.C. Watkins, "Crude Oil Prices Between 1985

and 1994: How Volatile inRelation to Other Commodities?" *Resource and Energy Economics*, Vol. 20, 1998, pp. 245-262.In general, Plourde and Watkins found that oil prices fluctuated more or at least as much as the most volatile of commodity prices; see the discussion in Hillard Huntington, "Energy Disruptions,Interfirm Price Effects, and the Aggregate Economy," Stanford Energy Modeling Forum,September 2002.

[114] International Energy Agency, "IEA Expresses Concern About High Oil Prices as it Celebratesits 30th Anniversary," Istanbul, April 2004; International Monetary Fund, *World Economic Outlook Report*, September 2003.

[115] Walter C. Labys, Globalization, Oil Price Volatility, and the U.S. Economy, 2001.

[116] Vincente Ramirez, "Oil Crises Delay – a World Oil Price Forecast," REXplore Zachasumsc,Switzerland, July 1999.

[117] John Schoen, "Oil Prices Include a Growing Risk Premium," Business with MSNBC, Oil andEnergy News, May 12, 2004.

[118] Jean-Marie Bourdaire, "Energy Supply Conditions and Oil Price Regime," presented at the Association for the Study of Peak Oil, Paris, May 2003.

[119] DOE EIA. "Long Term World Oil Supply." April 18, 2000.

[120] Deffeyes, K.S. Hubbert's Peak-The Impending World Oil Shortage. *Princeton University Press*. 2003. p. 134.

[121] This consensus emerged by the 1990s; see, for example, K. Lee, S. Ni, and R. Ratti, "OilShocks and the Macroeconomy: The Role of Price Variability," *Energy Journal*, Vol. 16, no. 4,1995.

[122] Once again, this experience may preclude such an option in the future, even though it may be called for. For example, by the 1990s, CBO had concluded that the threat posed by oil disruptions had declined; see U.S. Congressional Budget Office, op. cit.

[123] Estimates range from $2 trillion to more than $7 trillion (2004 dollars) --exclusive of military or political costs. See U.S. General Accounting Office, Energy Security: Evaluating U.S. Vulnerability To Oil Supply Disruptions and Options for Mitigating Their Effects, GAO/RCED-97-6,1997; David Greene and Nataliya Tishchishyna, *Cost* of Oil Dependence: A 2000 Update, OakRidge National Laboratory, May 2000; National Defense Council Foundation, The Hidden Cost of Imported Oil, October 2003.

[124] See the discussion in Roger Bezdek and John Taylor, "Allocating Petroleum Products During Oil Supply Disruptions," *Science*, June 19, 1981, Vol. 212, pp. 1357-1363.

[125] DOE, EIA Monthly Energy Review and Management Information Services, Inc., 2004

[126] In 1981, consumers spent nearly six percent of their incomes on gasoline, but in 2003 they spent only three percent of their incomes on gasoline; in 1985, gasoline and oil represented 20 percent of the cost of owning and operating a vehicle, but by 2002 represented only 10 percent of the cost.

[127] Between 1982 and 2002, the standard deviation in monthly oil prices was 29.5 percent of its mean, and the only other major commodity whose price exhibited similar volatility was coffee – 27.8 percent of its mean. See Andre Plourde and G.C. Watkins, "Crude Oil Prices Between 1985 and 1994: How Volatile in Relation to Other Commodities?" Resource and Energy Economics, Vol. 20, 1998, pp. 245-262. In general, Plourde and Watkins found that oil prices fluctuated more or at least much as the most volatile of commodity prices; see the discussion in Hillard Huntington, "Energy Disruptions, Interfirm Price Effects, and the Aggregate Economy," *Stanford Energy Modeling Forum*, September 2002.

[128] DOE EIA, *International Energy Outlook,* 2004.

[129] See the discussion in Roger H. Bezdek and Robert M. Wendling, "A Half-Century of Long-Range Energy Forecasts; Errors Made, Lessons Learned, and Implications for Forecasting,"

[130] Source: U.S. Department of Energy, Energy Information Administration, 2004.

[131] Management Information Services, Inc., and 20/20 Vision, Fuel Standards and Jobs: Economic, Employment, Energy, and Environmental Impacts of Increased CAFE Standards Through 2020, report prepared for the Energy Foundation, San Francisco, California, July 2002.

[132] Ibid.

[133] National Research Council, National Academy of Sciences, Effectiveness and Impact of Corporate Average Fuel Economy (CAFE) Standards, Washington, D.C.: *National Academy Press*, 2002.

[134] Ibid. Management Information Services, Inc., and 20/20 Vision, op. cit.; David L. Greene and John DeCicco, Engineering-Economic Analysis of Automotive Fuel Economy Potential in the United States, paper presented at the IEA International Workshop on Technologies to

Reduce Greenhouse Gas Emissions, Washington, D.C., May 1999; David Friedman, et al, Drilling in Detroit: Tapping Automaker Ingenuity to Build Safe and Efficient Automobiles, Union of Concerned Scientists, UCS Publications, Cambridge, MA, June 2001; Roland Hwang, Bryanna Millis, and Theo Spencer, Clean Getaway: Toward Safe and Efficient Vehicles, Natural Resources Defense Council: New York, July 2001; Brent D. Yacobucci, Sport Utility Vehicles, Mini-Vans and Light Trucks: An Overview of Fuel Economy and Emissions Standards, Congressional Research Service, U.S. Congress: Washington, D.C., (RS20298), January 16, 2001; Robert L Bamberger, Automobile and Light Truck Fuel Economy: Is CAFE Up to Standards? Washington, D.C.: Congressional Research Service, September 29, 2001; Energy and Environmental Analysis, Inc. Technology and Cost of Future Fuel Economy Improvements for Light-Duty Vehicles, prepared for the National Research Council, 2001.

[135] See Management Information Services, Inc., and 20/20 Vision, op. cit.; Roger H. Bezdek and Robert M. Wendling, "The Economic and Employment Effects of Increasing CAFE Standards." Energy Policy, 2004.

[136] U.S. Energy Information Administration, World Petroleum Consumption by Fuel database, 2003, and Oak Ridge National Laboratory, Transportation Energy Data Book, 2003. Japan has 10% of total vehicle registrations, Germany 9 percent, France 5 percent, and UK 5 percent, totaling (including the U.S.) 54 percent%. However, the U.S. has a higher miles per vehicle rate than any other developed country – it is less densely populated, has relatively inexpensive gasoline, and U.S. drivers do a large amount of discretionary driving.

[137] Kruger, P du P. "Startup Experience at Sasol's Two and Three." Sasol. 1983.

[138] Collings, J. "Mind Over Matter – The Sasol Story: A Half-Century of Technological Innovation," Sasol. 2002.

[139] DOE EIA. *International Energy Outlook*. 2004.

[140] "Sasol Taps Into China's Demand for Oil." Financial Times. July 8, 2004.

[141] Williams, B. "Heavy Hydrocarbons Playing Key Role in Peak Oil Debate, Future Supply."OGJ. July 28, 2003; DOE EIA. Early Release AEO 2004. December 16, 2003.

[142] "Drivers of the Energy Scene." *World Energy Council.* December 2003.

[143] Even under a crash program, 5 percent production increase in 10 years does not seem achievable, but roughly half that level might be possible. Our reasoning is strongly influenced by the need for relatively pure CO_2, which is difficult to obtain in most places around the world. This is especially true in the Middle East, where large sources of relatively pure CO_2 are somewhat rare at this time.

[144] McCaughey, J. "Is Gas OPEC in the Cards?" *Electricity Daily.* June 29, 2004.

[145] US DOE ONPOSR. *Strategic Significance of America's Oil Shale*, Vols I and II. March 2004.

[146] Rocky Mountain News, October 18, 2004, "Shale's New Hope: Shell Tests Technology to Cook Oil out of Rocks Underground," p. 1B.

INDEX

S

T

U

V

W

Y